An Introduction to the

Mathematical Structure
of
Quantum Mechanics

A Short Course for Mathematicians

ADVANCED SERIES IN MATHEMATICAL PHYSICS

I am grateful to the *Accademia dei Lincei* and the University of Rome "*La Sapienza*" for inviting me to give a series of lectures and for the opportunity of confronting this presentation of Quantum Mechanics with a very selected audience. The preparation of these lectures notes has benefited from enlightening discussions and collaboration with G. Morchio, to whom I am greatly indebted. I thank also the students of my courses for pointing out misprints or unclear points in the preliminary draft of the notes.

that the measurement of an observable in general limits the precision by which another observable can be subsequently measured. The mathematical abstraction of this deep physical fact is the realization that the algebra of observables is not described by an algebra of functions, but rather by an *algebra of operators in a Hilbert space*. As mentioned before, the passage from the commutative structure of classical mechanics and/or of classical statistical mechanics to the non-commutative structure of quantum mechanics is the deep and crucial feature shared by the modern non-commutative extension of calculus, probability, geometry etc.

Last but not least, quantum mechanics had a dramatic impact on the development of mathematical logic, giving rise to the so-called *Quantum Logic* : whereas the lattice of propositions of classical logic has the structure of a Boolean algebra (equivalently that of a lattice of commutative projections), the lattice of quantum propositions is non-boolean and it corresponds to a lattice of non-commutative projections [Birkhoff and Von Neumann (1936); Beltrametti and Cassinelli (1981); Cohen (1989); Garden (1984); Hooker (1975); Pitowski (1989); Rèdei (1998)].

The aim of these lectures is to provide at least the flavor of the philosophical revolution induced by quantum mechanics concerning the mathematical description of physical systems. The lectures are primarily addressed to people interested in questions of principle and in the mathematical foundations of physical theories, also in view of the fertile mutual influence between theoretical physics and mathematics.

In order to make the ideas at the basis of quantum mechanics understandable also to people with a mathematical education but with no great familiarity with physics, we will reduce the detailed description of the many experimental facts which led to the crisis of classical mechanics to the minimum and will rather extract and emphasize the overall simple and profound message for the mathematical description of quantum systems.

Once the Heisenberg revolutionary discovery has been accepted, namely that there are intrinsic limitations to the precise measurements of physical quantities (Heisenberg's uncertainty relations) leading to the non abelianity of the algebra of observables, the whole mathematical structure of Quantum Mechanics follows as a theorem (Gelfand-Naimark): the states of a physical system are described by vectors of a Hilbert space and the observables by Hilbert space operators. Also the Schroedinger formulation of Quantum Mechanics in terms of wave functions follows from Von Neumann uniqueness theorem on the regular (irreducible) representations of the Weyl algebra.

Those who will hopefully find the subject sufficiently interesting and stimulating are warmly referred to standard textbooks to deepen the mathematical and logical structure of quantum mechanics and to appreciate its impact on the description of the physical world [Dirac (1958); Feynman et al.(1963); Heisenberg (1930); Jauch (1968); Von Neumann (1955); Mackey (1963); Piron (1976); Segal (1963)].

GENERAL REFERENCES

J. Dixmier, *Von Neumann Algebras*, North-Holland 1981

J. Dixmier, *C* Algebras*, North-Holland 1977

R.V. Kadison and J.R. Ringrose, *Fundamentals of the Theory of Operator Algebras*, Vol.I-IV, Academic Press 1983

S. Stratila and L. Zsido', *Lectures on Von Neumann Algebras*, Abacus Press 1979

M. Takesaki, *Theory of Operator Algebras*, Vol.I, Springer 1979

P.E.T. Jorgensen and P.S. Muhly eds., *Contemporary Mathematics Vol. 62, Operator Algebras and Mathematical Physics*, Am. Math. Soc. 1987

Proc. Symposia in Pure Mathematics, *Operator Algebras and Applications*, Vol.I,II, Am. Math. Soc. 1982

O. Bratteli and D.W. Robinson, *Operator Algebras and Quantum Statistical Mechanics*, Vol.I, Springer 1979; Vol.II, Springer 1981

G.B. Folland, *Harmonic Analysis in Phase Space*, Princeton Univ. Press 1989

L. Garding and A.S. Wightman, Proc. Natl. Acad. Sci. USA, **40**, 617 (1954)

C.R. Putnam, *Commutation Properties of Hilbert Space Operators and Related Topics*, Springer 1967

H.L. Cycon et al., *Schroedinger Operators*, Springer 1987

M.S.P. Eastham and H. Kalf, *Schroedinger type operators with continuum spectra*, Pitman 1982

S. Graffi, *Schroedinger Operators*, CIME course 1984, Lect. Notes Math. 1159, Springer 1985

P.D. Hislop and I.M. Sigal, *Introduction to Spectral Theory with Applications to Schroedinger Operators*, Springer 1996

M. Schechter, *Operator Methods in Quantum Mechanivs*, North-Holland 1981

W.O. Amrein et al., *Scattering Theory in Quantum Mechanics. Physical Principles and Mathematical Methods*, W.A. Benjamin 1977

J.A. La Vita and J.P. Marchand eds., *Scattering Theory in Mathematical Physics*, D. Reidel 1974

P.D. Lax and R.S. Phillips, *Scattering Theory*, Academic Press 1989

P.D. Lax and R.S. Phillips, *Scattering Theory for Automorphic Functions*, Princeton Univ. Press (Ann. Math. Studies) 1976

D.B. Pearson, *Quantum Scattering and Spectral Theory*, Academic Press 1988

P.A. Perry, *Scattering Theory by Enss Method*, Chur (Math. Reports) 1983

I.M. Sigal, *Scattering Theory for Quantum Mechanical Systems: Rigorous Results*, Lect. Notes Math. 1011, Springer 1983

D.R. Yafaev, *Mathematical Scattering Theory*, Am. Math. Soc. 1992

Ph. Blanchard et al., *Mathematical and Physical Aspects of Stochastic Mechanics*, Springer 1987

K.L. Chung and R.J. Williams, *Introduction to Stochastic Integration*, Birkhauser 1982

J. Glimm and A. Jaffe, *Quantum Physics. A Functional Integral Point of View*, Springer 1987

M. Kac, *Integration in Function Spaces and Some of Its Applications*, Lezioni Fermiane, Scuola Normale Superiore Pisa 1980

G. Roepstorff, *Path Integral Approach to Quantum Physics. An Introduction*, Springer 1993

B. Simon, *Functional Integration and Quantum Physics*, Academic Press 1979

H. Araki, *Quantum and Non-Commutative Analysis: Past, Present and Future Perspectives*, Kluwer 1995

A. Connes, *Géométrie non commutative*, InterEdition Paris 1990; *Non-Commutative Geometry*, Academic Press 1994

A. Connes, *Mathématiques Quantiques. Geometrie Non-Commutative et Physique Quantique*, Soc. Math. France 1992

A. Connes, *Quantized Calculus and Applications*, in Proc. XIth Int. Cong. of Mathematical Physics, D. Jagolnitzer ed., International Press 1995, p.15; Jour. Math. Phys. **36**, 6194 (1995)

A. Connes, *Non-Commutative Geometry and Physics*, Les Houches Lectures 1992, in *Gravitation and Quantizations*, B. Julia and J. Zinn-Justin eds. North-Holland 1995

C. Kassel, *Quantum Groups*, Springer 1995

A.A. Kirillov, *Representation Theory and Non-Commutative Harmonic Analysis*, Encyclopedia Math. Sci. N. 59, Springer 1995

J. Madore, *Non-Commutative Differential Geometry and Its Physical Applications*, London Math. Soc. Lect. Notes 206, Cambridge Univ. Press 1995

Y.I. Manin, *Quantum Groups and Non-Commutative Geometry*, CRM 1988

P.-A. Meyer, *Quantum Probability for Probabilists*, Springer 1992

K.R. Parthasarathy, *An Introduction to Quantum Stochastic Calculus*, Birkhäuser 1992

P. Biane, in *Lectures on Probability Theory*, Lect. Notes Math. 1608, Springer 1995

G. Birkhoff and J. von Neumann, Ann. Math. **37**, 823 (1936)

E.G. Beltrametti and G. Cassinelli, *The Logic of Quantum Mechanics*, Addison Wesley 1981

D.W. Cohen, *An Introduction to Hilbert Space and Quantum Logic*, Springer 1989

R.W. Garden, *Modern Logic and Quantum Mechanics*, A. Hilger 1984

C.A. Hooker ed., *The Logico-Algebraic Approach to Quantum Mechanics*, D. Reidel 1975

I. Pitowsky, *Quantum Probability and Quantum Logic*, Lect. Notes Phys. 321, Springer 1989

M. Rédei, *Quantum Logic in Algebraic Approach*, Kluwer 1998

P.A.M. Dirac, *The Principles of Quantum Mechanics*, Oxford Claredon Press 1958

R.P. Feynman et al., *The Feynman Lectures on Physics*, Vol.III, Addison Wesley 1963

W. Heisenberg, *The Physical Principles of the Quantum Theory*, Dover 1930

J.M. Jauch, *Foundations of Quantum Mechanics*, Addison Wesley 1968

J. Von Neumann, *Mathematical Foundations of Quantum mechanics*, Princeton University Press 1955

G.W. Mackey, *Mathematical Foundations of Quantum Mechanics*, W.A. Benjamin 1963

C. Piron, *Foundations of Quantum Physics*, W.A. Benjamin 1976

I.E. Segal, Ann. Math. (2), **48**, 930 (1947); *Mathematical Problems of Relativistic Physics*, Am. Math. Soc. 1963

Chapter 1

Mathematical description
of a physical system

1.1 Atomic physics and the crisis of classical mechanics

Quantum mechanics was invented on the basis of very cogent physical reasons. The large body of physical motivations and experimental facts, which are usually regarded as convincing enough by physicists, may appear not sufficiently forcing to mathematicians, also in view of the fact that the philosophical change involved is rather dramatic. To make the message more direct we will only briefly recall the basic experimental facts, which led to the crisis of classical mechanics for the description of atomic physics; we will rather dwell on the logical consequences of Heisenberg uncertainty relations and on the new mathematical structures which follow from them.

We list some of the most important phenomena in conflict with classical physics:

1) **Atomic physics**. There is a host of experimental evidence that an atom consists of a nucleus, made of neutrons and protons, and of electrons bound to it in a sort of planetary system, with the Coulomb potential playing the role of the gravitational potential. For example, the hydrogen atom has a nucleus made of a proton, of positive charge e, and an electron (of negative charge $-e$); the mass of the proton is about 1800 times the mass of the electron $m_e \simeq 10^{-27}$ gr. Such a planetary picture, which on one side is strongly supported by experimental data (typically Rutherford's experiment with α particles) leads to the following classical paradoxes:

i) all atoms have approximately the same dimensions $\simeq 10^{-8}$ cm., whereas classically the dimension of the orbit of a planet varies with the energy

(which can be arbitrary)

ii) the atoms are stable and therefore so must be the electron orbits, incompatibly with the laws of electrodynamics, according to which an accelerated charge emits electromagnetic radiation with inevitable energy loss. The electron should therefore collapse on the nucleus and a simple calculation shows that correspondingly the lifetime of an orbit of dimension 10^{-8} cm. should be of 10^{-10} sec.; in this case the dimensions of an atom would rapidly become those of its nucleus, namely of the order of 10^{-13} cm.

iii) the spectrum of the radiation absorbed or emitted by an atom, under the influence of external forces, consists of a discrete series of frequencies, contrary to the laws of classical physics, according to which the frequency ν of a planetary motion and therefore the radiation spectrum varies continuously with the dimensions of the orbit, for example for a circular orbit of radius r

$$\nu = (2\pi)^{-1} \sqrt{\frac{e^2}{m}}\, r^{-\frac{3}{2}}. \tag{1.1.1}$$

All this suggests that only a discrete set of orbits is allowed, equivalently that only a discrete set of frequencies of the electron (periodic) motion are allowed (*quantization of the electron periodic motion*).

2) **Photoelectric effect**. If light of short wavelength is sent on a metallic surface, electrons are emitted (roughly they gain enough energy to escape from being bounded inside the metal). Classically, one would expect that the phenomenon is crucially governed by the energy carried by the electromagnetic radiation, i.e. by the light intensity. On the contrary, what happens is that the crucial quantity is the frequency ν of the electromagnetic wave; indeed

i) the electron emission occurs only if $\nu > \nu_0 = $ *frequency threshold* (which depends on the metal),

ii) the maximum kinetic energy T of the emitted electrons is a function of the frequency,

$$T = a(\nu - \nu_0),$$

a a positive constant, rather than a function of the intensity of the radiation,

iii) the effect of the intensity (for fixed frequency ν) is only that of increasing the number of emitted electrons (not their energy !).

As argued by Einstein, this suggests that at the microscopic level the electromagnetic radiation of frequency ν does not carry energy in a continuous way, proportional to the radiation intensity, but rather in discrete fractions called *quanta* or *photons* each with energy

$$E = h\nu, \tag{1.1.2}$$

where $h = 6.6\ 10^{-27}$ erg sec is the Planck constant (and with momentum $p = h\nu/c$), and that the probability of more than one photon absorption by one electron is depressed. For fixed frequency, the intensity of the radiation is proportional to the number of photons carried by it and not to the energy of each photon (*quantum character of the electromagnetic radiation*). In this case, this can be interpreted as the evidence for the *particle or corpuscular behaviour* of *light* at the microscopic level.

3) **Particle-Wave duality of matter**. The above corpuscular behaviour of light (related to the way energy is carried by e.m. radiation) does not require a corpuscular localization of photons. In fact, in general they are not strictly localized in space; they are described by "wave functions" and this explains why interference and diffraction phenomena characterize the electromagnetic waves (*particle-wave duality of photons*). For example, in Bragg's experiment, if one sends a beam of light of wavelength λ on the plane surface of a crystal made of lattice planes at distance d, then, if θ is the incidence angle of the light beam on the crystal surface, the optical paths of two rays reflected by two lattice planes at distance d differ by the amount $2d \sin \theta$. Therefore, one has constructive interference for the reflected rays if $2d \sin \theta = n \lambda$, $n \in \mathbf{N}$.

A similar experiment was done by Davisson and Germer replacing the light beam by a well focused beam of electrons, all of (approximately) the same energy E. The result was a constructive interference if $2d \sin \theta = nh/p$, where $p = \sqrt{2mE}$ is the electron momentum. This indicates a wave behaviour of matter with wave length λ given by the De Broglie relation

$$\lambda = h/p = h/\sqrt{2mE} \qquad (1.1.3)$$

(*particle-wave duality of matter*).

There are other important experimental facts which played a relevant role in the birth of Quantum Mechanics, like the *black-body radiation* and the temperature dependence of the *specific heats of gases*, but their discussion would lead us too far.[1]

The above sketchy account of the experimental facts, which led to the crisis of Classical Mechanics, may not provide a convincing evidence for the need of radical changes, especially if one is not familiar with the sharp constraints of classical physics. The implications of the above experiments at the level of general strategies may not appear logically inevitable, but a critical analysis would actually show that there is no alternative to changing the roots of classical physics.[2] Regrettably, we have to omit a detailed

---

[1] A discussion of the experiments at the basis of quantum mechanics can be found in many textbooks, see e.g. A. Messiah, *Quantum Mechanics*, North-Holland 1961 Vol. I, Chaps. I-III; S-I. Tomonaga, *Quantum Mechanics*, North-Holland 1962, Vol. I, Chap. 1,2; M. Born, *Atomic Physics*, Blakie 1958, Chap. VIII, Sects. 1-3.

[2] See e.g. J.A. Wheeler and W.H. Zurek, *Quantum Theory and Measurement*, Princeton University Press, 1983; A. Peres, *Quantum Theory: Concepts and Methods*, Kluwer 1993.

discussion of these points, also because it would rely on a non-superficial mastering of physical arguments. Rather, we will follow the simpler logic of showing that the foundations of quantum mechanics can be deduced essentially by a single crucial fact, namely the *Heisenberg's realization of the uncertainty relations*, which affect the measurement of physical quantities at the microscopic level.

To fully appreciate the strength of Heisenberg intuition, we will start by a general revisitation of the mathematical structures and ideas underlying the foundations of classical mechanics.

1.2   Mathematical description of classical Hamiltonian systems

In order to realize the roots of the conflict between atomic physics and classical physics, we will isolate the basic structure underlying the mathematical description of a classical mechanical system.

Kinematics. The *configuration* or the *state* of a *classical Hamiltonian system* is (assumed to be) described by a set of canonical variables $\{q, p\}$, $q = (q_1, ..., q_n)$, $p = (p_1, ..., p_n)$, briefly by a point $P = \{q, p\} \in \Gamma \equiv$ phase space manifold. For simplicity, in the following, we will confine our discussion to the case in which Γ is compact. This is, e.g., the case in which the system is confined in a bounded region of space and the energy is bounded.

The physical quantities or *observables* of the system, clearly include the q's and p's and their polynomials and therefore, without loss of generality, we can consider as observables their sup-norm closure, i.e. (real) continuous functions $f(q, p) \in C_{\mathbf{R}}(\Gamma)$, (for a further extension see the remark after eq. (1.2.3)).

Every state P determines the values of the observables on that state and conversely, by the Stone-Weierstrass and Urysohn theorems, any $P \in \Gamma$ is uniquely determined by the values of all the observables on it (duality relation between states and observables).

Dynamics. The relation between the measurement of an observable f at an initial time t_0 and at any subsequent time t is given by the time evolution of the canonical variables

$$q \to q_t = q(t, q, p), \quad p \to p_t = p(t, q, p), \quad f_t(q, p) \equiv f(q_t, p_t). \quad (1.2.1)$$

The time evolution of the canonical variables is given by the Hamilton equations

$$\dot{q} = \frac{\partial H}{\partial p}, \quad \dot{p} = -\frac{\partial H}{\partial q}, \quad (1.2.2)$$

where $H = H(q, p)$ is the Hamiltonian. Under general conditions (typically if grad$H$ is Lipschitz continuous), for any initial data, the above system of equations has a unique solution local in time, which can be extended to all times under general conditions, e.g. if the surfaces of constant energy are compact [3].

The mathematical structures involved are the theory of functions (on phase space manifolds) and the theory of first order differential equations, defined by Hamiltonian flows on phase space manifolds.

From the above picture of elementary Hamiltonian mechanics one can extract the following algebraic structure.

I. **Algebraic properties of the classical observables**. The observable quantities associated to a classical system generate an abelian algebra $\mathcal{A}$ of real or more generally complex [4] continuous functions on the (compact) phase space (the product being given by the pointwise composition of functions $(fg)(x) = f(x)g(x)$ etc.). This algebra has an identity **1** given by the function $f = 1$ and a natural involution or $\star$ operation is defined by the ordinary complex conjugation, $f^\star(x) = \bar{f}(x)$, so that $\mathcal{A}$ is a $\star$-algebra with identity. To each element $f \in \mathcal{A}$ one can assign a norm, $\| f \|$, given by the sup-norm

$$\| f \| = \sup_{x \in \Gamma} | f(x) |, \tag{1.2.3}$$

so that $\mathcal{A}$ is a Banach space with respect to this norm. The product is continuous with respect to the norm topology since

$$\| fg \| \leq \| f \| \, \| g \| \tag{1.2.4}$$

and therefore $\mathcal{A}$ is a *Banach $\star$-algebra*. Finally, the following property holds (C^*-*condition*)

$$\| f^\star f \| = \| f \|^2. \tag{1.2.5}$$

Technically, an algebra with the above properties is called an *abelian C^*-algebra.*[5]

II. **States as linear functionals**. From an operational point of view, the identification of the states of a classical system with points of the phase space Γ relies on the unrealistic idealization, according to which the configuration of the system is sharply defined by measuring the canonical variables (typically positions and velocities) with infinite precision. Clearly, from a

---

[3]For a discussion of the existence theorems see V. Arnold, *Ordinary Differential Equations*, Springer 1992, and V. Arnold, *Mathematical Methods of Classical Mechanics*, Springer 1989

[4]Such extension is both natural and convenient and in any case completely determined by the real subalgebra.

[5]For a beautiful account of the theory of abelian C^*-algebras see I.M. Gelfand, D.A. Raikov and G.E. Shilov, *Commutative Normed Rings*, Chelsea 1964; see also R.S. Doran and V.A. Belfi, *Characterization of C^*-Algebras*, Dekker 1986, esp. Chap. 2.

physical point of view it is more sensible to admit that in the preparation or detection of a state of a physical system a certain undeterminacy is unavoidable so that the configuration of the system at the initial time t_0 is known within a certain error, which inevitably propagates in time.

Since a state of the system is characterized by the measurements of the observables in that state, it is convenient to recall the operational meaning of such a procedure.

As it is well known, measurements with infinite precision are not possible and therefore the standard experimental way of associating a value of an observable f to a state ω is to perform replicated measurements of f, $m_1^{(\omega)}(f)$, $m_2^{(\omega)}(f)$, ..., $m_n^{(\omega)}(f)$, on the system in the given state ω or more generally on replicas of it and to compute the average

$$< f >_n^{(\omega)} \equiv [m_1^{(\omega)}(f) + m_2^{(\omega)}(f) + m_n^{(\omega)}(f)]/n.$$

The limit $n \to \infty$ (whose existence is part of the foundations of experimental physics) defines the *expectation of f* on the state ω

$$\omega(f) \equiv \lim_{n \to \infty} < f >_n^{(\omega)} \tag{1.2.6}$$

as *average of the results of measurements of f* in the state ω.

The coarseness affecting the measurements of f is given by

$$(\Delta_\omega f)^2 \equiv \omega((f - \omega(f))^2), \tag{1.2.7}$$

since it indicates how much the results of measurements of f in the given state ω depart from the average $\omega(f)$; it is also called the mean square deviation or the variance of f (relative to ω). More generally, all experimental information on the measurement of an observable f in the state ω are recorded in the expectations of the polynomials of f. [6]

This is the way the experimental results are recorded and the operational identification of a state of a physical system is therefore given by the set of expectation values of its observables. Since the expectation $\omega(f)$ of an observable f has the interpretation (and actually corresponds to the operational definition) of the average of the results of the measurements of f in the given state ω, it follows that such expectations are linear, i.e.

$$\omega(\lambda f_1 + \mu f_2) = \lambda \omega(f_1) + \mu \omega(f_2), \quad \forall f_1, f_2 \in \mathcal{A}, \ \lambda, \mu \in \mathbf{C} \tag{1.2.8}$$

and satisfy the *positivity condition*, namely

$$\omega(f^* f) \geq 0, \quad \forall f \in \mathcal{A}. \tag{1.2.9}$$

---

[6]Such characterization of the measurements of f is somewhat related to the moment problem, for which a bound on the expectations of the polynomials of f is provided by the scale bound of the experimental apparatus associated to the measurements of f (such strict relations between observables and apparatuses yielding their measurements will be further discussed and exploited in the next section).

The positivity condition $(\omega((A+B)^*(A+B)) \geq 0)$ implies the validity of Cauchy-Schwarz' inequality

$$|\omega(A^*B)| \leq \omega(A^*A)^{1/2}\,\omega(B^*B)^{1/2}, \quad \forall A, B \in \mathcal{A}, \tag{1.2.10}$$

and therefore $\omega(\mathbf{1}) > 0$ unless ω is the trivial state ($\omega = 0$ on $\mathcal{A}$). Thus, without loss of generality, given a (non-trivial) state ω one may always normalize it: $\omega \to \omega_{norm} = \omega(\mathbf{1})^{-1}\omega$, so that $\omega_{norm}(\mathbf{1}) = 1$.

Thus, in conclusion and quite generally, a classical system is defined by the abelian C^*-algebra $\mathcal{A}$ of its observables and a *state* of a classical system is a *normalized positive linear functional* ω on $\mathcal{A}$. A state ω on a C^*-algebra of continuous functions $C(X)$ on a compact (Hausdorff) space X is automatically continuous and therefore by the Riesz-Markov representation theorem [7] it defines a unique (regular Borel) measure μ_ω on X such that

$$\omega(f) = \int_X f\, d\mu_\omega, \quad \mu_\omega(X) = \omega(\mathbf{1}) = 1, \tag{1.2.11}$$

so that the expectations have a probabilistic interpretation [8]. Thus, the operational characterization of a state of a physical system leads to its description by a probability distribution rather than by a point of the phase space and the observables get the meaning of *random variables.*

The above considerations support the description of observables by continuous functions and justify the possible extension of the concept of observable to the pointwise limits of continuous functions, almost everywhere with respect to μ_ω.

Clearly, the above general concept of state contains as a very special case the definition of state in elementary mechanics; in fact, if μ_{P_0} is the probability measure concentrated on the point $P_0 = \{q_0, p_0\}$, i.e. for any measurable set S, $\mu_{P_0}(S) = 1$ if $P_0 \in S$ and $= 0$ otherwise, (namely μ_{P_0} is a Dirac δ function), then the corresponding state ω_{P_0} is

$$\omega_{P_0}(f) = \int_\Gamma f\, d\mu_{P_0} = f(P_0), \tag{1.2.12}$$

i.e. it is described by the point P_0. Such states are also called *pure states* since they cannot be written as convex linear combinations of other states (see also the next Sections). Clearly, for a pure state ω the variance vanishes, i.e. f takes a sharp value f_ω in the state ω, in the sense that all the measurements yield the same result, $f_\omega = \omega(f)$, i.e. there is no dispersion (such states are also called *dispersion free states*).

---

[7] A simple discussion is in M. Reed and B. Simon, *Methods of Modern Mathematical Physics*, Academic Press, Vol. I (Functional Analysis), Chap. IV, Sect. 4.

[8] For an introduction to the theory of probability and in particular to the concept of random variable see, e.g. J. Lamperti, *Probability*, W.A.Benjamin 1966, esp. Chap. 1, and H.G. Tucker, *A Graduate Course in Probability*, Academic Press 1967, esp. Chaps. 1,2.

For the general states defined above the time evolution can be defined by duality in terms of the time evolution of the observables

$$\omega_t(f) \equiv \omega(f_t). \tag{1.2.13}$$

The above mathematical description of a state of a classical system is not only strongly suggested by operational arguments, concerning the measurement of physical quantities, but it is absolutely necessary for the mathematical and physical description of complex systems, i.e. when the number of degrees of freedom become very large, typically 10^{23} for thermodynamical systems. In this case, it is technically impossible to control an initial value problem for such a large number of variables and it is also physically unreasonable to measure all of them. Moreover, such idealistic description of a complex system is not what is required on physical grounds to account for the time evolution of physically realizable measurements. Thus, the mathematical and physical description of a complex system inevitably requires new mathematical ideas and structures with respect to those of classical analysis, namely the *theory of random variables.* This was indeed the revolutionary step taken by Boltzmann in laying the foundations of Classical Statistical Mechanics and in deriving classical thermodynamics from the mechanical properties of complex systems.

III. **Algebraic Dynamics.** Under general regularity conditions, in the concrete case of the canonical realization of a classical system, the time evolution $\{q, p\} \rightarrow \{q_t, p_t\}$ is continuous in time $t \in \mathbf{R}$, with a continuous dependence on the initial data at time t_0 and therefore it defines a one-parameter family of continuous invertible mappings $\alpha_{t_0, t}$ of $C(\Gamma)$ into itself,[9] which preserves all the algebraic relations, including the $*$-operation (by eq. (1.2.2) $\alpha_t(f\, g) = \alpha_t(f)\, \alpha_t(g)$, $\alpha_t(f^*) = (\alpha_t(f))^*$). A linear invertible mapping of a C^*-algebra into itself, which preserves the algebraic relations is called a $*$-automorphism (it follows from a general result that it necessarily preserves the norm, see e.g. Proposition 2.2.3 in the next chapter).

Quite generally, given an abelian C^*-algebra $\mathcal{A}$ of observables a time-translation invariant (reversible) dynamics, (i.e. one which depends only on the difference $t - t_0$), can be algebraically defined as a one-parameter group of $*$-automorphisms α_t of $\mathcal{A}$, $t \in \mathbf{R}$ and by duality one can define a one-parameter group of transformations α_t^* of states into states given by

$$\omega_t(A) \equiv (\alpha_t^* \omega)(A) \equiv \omega(\alpha_t(A)), \tag{1.2.14}$$

---

[9]In the case of non regular dynamics, i.e. when the algebra of continuous functions on Γ is not stable under time evolution, one has to identify the observables with a larger C^*-algebra, than that of the continuous functions on Γ, in order to guarantee stability under time evolution (a necessary requirement for a reasonable physical interpretation).

(*time evolution of the states*). The abstract version of the continuity in time is that for any state ω, $\omega(\alpha_t(A))$, is continuous in time $\forall A \in \mathcal{A}$; technically α_t is said to be *weakly continuous*.

The recognition of the above mathematical structure at the basis of the standard description of classical systems suggests an abstract characterization of a classical (Hamiltonian) system, with no a priori reference to the explicit realization in terms of canonical variables, phase space, continuous functions on the phase space, etc. In this perspective, since a physical system is described in terms of measurements of its observables, one may take the point of view that a classical system is *defined* by its physical properties, i.e. by the algebraic structure of the set of its measurable quantities or observables, which generate an abstract abelian C^*-algebra $\mathcal{A}$ with identity. The states of the system being fully characterized by the expectations of the observables are described by normalized positive linear functionals on $\mathcal{A}$ and the dynamics is a one-parameter group of *-automorphisms of $\mathcal{A}$.

It is important to mention that quite generally, by the Gelfand-Naimark representation theorem [10], an (abstract) abelian C^*-algebra $\mathcal{A}$ (with identity) is isometrically isomorphic to the algebra of complex continuous functions $C(X)$ on a compact Hausdorff topological space X, where X is intrinsically defined as the Gelfand spectrum of $\mathcal{A}$.

From the point of view of general philosophy, the picture emerging from the Gelfand theory of abelian C^*-algebras has far reaching consequences and it leads to a rather drastic change of perspective. In the standard description of a physical system the geometry comes first: one first specifies the coordinate space, (more generally a manifold or a Hausdorff topological space), which yields the geometrical description of the system, and *then* one considers the abelian algebra of continuous functions on that space. By the Gelfand theory the relation can be completely reversed: one may start from the abstract abelian C^*-algebra, which in the physical applications may be the abstract characterization of the observables, in the sense that it encodes the relations between the physical quantities of the system, and then one reconstructs the Hausdorff space such that the given C^*-algebra can be seen as the C^*-algebra of continuous functions on it. In this perspective, one may say that the algebra comes first, the geometry comes later. The total equivalence between the two possible points of view indicates a purely algebraic approach to geometry: compact Hausdorff spaces are described by abelian C^*-algebras with identity, whereas if the algebra does not have an identity one has a locally compact Hausdorff space.

Non-commutative geometry is the structure emerging when the algebra is non-commutative. [11]

---

[10] For the convenience of the reader a brief outline of the Gelfand-Naimark theory is given in Appendix B.

[11] A. Connes, *Non Commutative Geometry*, Academic Press 1994.

1.3   General mathematical description of a physical system

In this section we argue that the structure of C^*-algebra of observables and states is the suitable language for the mathematical description of a physical system in general (including the atomic systems), with no reference to classical mechanics and its standard paradigms. [12]

I. **Observables.** From an operational point of view, a physical system is defined by its physical properties, i.e. by the set $\mathcal{O}$ of the physical quantities (briefly called *observables*) which can be measured on it and by the relations between them. Each observable has to be understood as characterized by a concrete physical apparatus yielding its measurements.[13]

For any $A \in \mathcal{O}$ and $\lambda \in \mathbf{R}$, one can define the observable λA as the observable measured by rescaling the apparatus by λ. By similar considerations one justifies the existence of elementary functions of an observable like the powers (with the standard elementary properties): if $A \in \mathcal{O}$, A^2 may be interpreted as the observable associated with squaring the apparatus scale (equivalently by squaring the results of measurements). Similarly, one defines the powers A^m, and their products $A^m A^n \equiv A^{m+n}$. It follows from this definition that A^0 is the observable whose results of measurements always take the value 1, independently of the state on which the measurement is done.

In the same way, one defines a polynomial of A as the observable obtained by taking as the new apparatus scale the given polynomial function of the scale for A. An element $A \in \mathcal{O}$ is said to be *positive* if all the *results of measurements* of A are positive numbers. By the operational definition of elementary functions of A, this implies that (and it is actually equivalent to) A is of the form $A = B^2$, $B \in \mathcal{O}$.

II. **States.** A *state ω of a physical system* is characterized by the results of the measurements of the observables in the sense that the *average over the results of measurements* of an observable A, when the system is in a state ω, defines the *expectation $\omega(A)$* and the state ω is completely characterized by

---

[12] The C^*-algebraic approach to classical and quantum physics has been pioneered by I. Segal, Ann. Math.(2) **48**, 930 (1947); *Mathematical Problems of Relativistic Physics*, Am. Math. Soc. 1963; see also P. Jordan, J. Von Neumann and E.P. Wigner, Ann. Math. **35**, 29 (1934) for an early proposal and G.G. Emch, *Algebraic Methods in Statistical Mechanics and Quantum Field Theory*, Wiley-Interscience 1972, esp. Chap. 2, for a historical review and a systematic treatment.

[13] The possibility that two distinct experimental apparatuses effectively define the same observable will be discussed below. It is convenient to deal with dimensionless observables, whose measurements are defined as ratios with respect to a set of reference measurements (e.g. of length, mass etc.); such a choice of scale is implicit in each physical apparatus.

all its expectations $\omega(A)$, when A varies over $\mathcal{O}$; thus ω is a (real) *functional* on $\mathcal{O}$. By the operational definition of $\omega(A)$ it easily follows that ω is a homogeneous functional

$$\omega\left(\lambda A\right) \ = \ \lambda\,\omega(A) \ , \quad \forall \lambda \in \mathbf{R}, \tag{1.3.1}$$

and that $\omega\left(A^{n} + A^{m}\right) = \omega\left(A^{n}\right) + \omega\left(A^{m}\right)$.

The realization that the only operational way of characterizing a state is in terms of its expectations of the observables, requires that two states yielding the same expectations must be identified, i.e.

$$\omega_1(A) = \omega_2(A), \quad \forall A \in \mathcal{O} \ , \tag{1.3.2}$$

must imply $\omega_1 = \omega_2$, (briefly the observables separate the states).

On the other hand, if we put at the basis of the mathematical description of a physical system the fact that the experimental way of identifying an observable is in terms of its expectations on the states, then two observables A and B having the same expectations on all the states, $\omega(A) = \omega(B)$, $\forall \omega$, cannot be distinguished. Such property of the states, of completely characterizing the observables and their relations, can be viewed as a *completeness of the states* with respect to the observables. This means that the states define an equivalence relation, denoted by $\sim$, between the elements of $\mathcal{O}$: $A \sim B$, if $\omega(A) = \omega(B)$ for all the physical states ω , (for example two distinct experimental apparatuses may effectively define the same observable). Two equivalent observables must therefore be identified and in the following the set $\mathcal{O}$ of observables will always denote the corresponding set of equivalence classes.

The definition of the zeroth power of an observable A implies that for any physical state $\omega\left(A^0\right) = 1$, since all the results of measurements take the value 1 and therefore so does their average. Thus, all the zeroth powers of observables fall in the same equivalence class which will be called the identity and denoted by $\mathbf{1}$. Clearly, for any state

$$\omega\left(\mathbf{1}\right) = 1, \tag{1.3.3}$$

i.e. a state of a physical system is a *normalized functional* on $\mathcal{O}$.

From the existence of products of powers of an observable follows the existence of a product of $\mathbf{1}$ with any observable:

$$A\,\mathbf{1} = \ A\,A^0 = \ A = \ \mathbf{1}A.$$

As we have also seen in the previous section, a crucial *positivity* property must be satisfied by the states. Since the expectation $\omega(A)$ is the *average* over the results of measurements in the state ω , it follows that for any state ω , if A is positive, then

$$\omega\left(A\right) = \omega\left(B^2\right) \geq 0, \tag{1.3.4}$$

i.e. ω is a *normalized positive functional* on $\mathcal{O}$.

By the completeness property of the states in identifying the observables, all the properties of an observable A have to be encoded in its expectation values and therefore, in particular, the positivity of an observable A has to be equivalently described by the positivity of all its expectations, $\omega(A) \geq 0$ for all ω.

III. **C^*-algebraic structure.** Since, as discussed above, an observable A is defined in terms of a concrete experimental apparatus, which yields the numerical results of measurements in any state, and since each concrete experimental apparatus has inevitable limitations implying a scale bound independent of the state on which the measurement is performed, the results of measurements of A in the various states is a bounded set of numbers, with bound related to the scale bound of the associated experimental apparatus. To each observable A it is then natural to associate the finite bound

$$||A|| \equiv \sup_{\omega} |\omega(A)| < \infty. \tag{1.3.5}$$

Clearly, by the homogeneity of the states

$$||\lambda A|| = |\lambda| \, ||A||, \quad \forall \lambda \in \mathbf{R}. \tag{1.3.6}$$

Moreover, since the states separate the observables, $||A|| = 0$ implies $A = 0$. From the definition of $||A||$ it follows that

$$||A^2|| = ||A||^2. \tag{1.3.7}$$

In fact, by definition, for any physical state ω, $\omega(||A||\mathbf{1} \pm A) \geq 0$, so that $||A||\mathbf{1} \pm A$ are both positive; then $(||A||\mathbf{1} - A)(||A||\mathbf{1} + A)$ is a positive polynomial of A and

$$||A||^2 - \omega(A^2) = \omega((||A||\mathbf{1} - A)(||A||\mathbf{1} + A)) \geq 0, \forall \omega \tag{1.3.8}$$

which implies $||A||^2 \geq ||A^2||$. On the other side, the positivity of

$$(||A||\mathbf{1} \pm A)^2 = ||A||^2 + A^2 \pm 2||A||A$$

implies that for any state ω

$$2||A|| \, |\omega(A)| \leq ||A||^2 + \omega(A^2) \leq ||A||^2 + ||A^2|| \tag{1.3.9}$$

and therefore $||A||^2 \leq ||A^2||$.

The duality relation between observables and states allows to display and define linear structures in $\mathcal{O}$. We have already argued that the sum of polynomials of one observable has a well defined operational meaning;

actually for a larger class of pairs A, B, (e.g. the kinetic and the potential energy) the sum can be defined in the sense that there is an observable C such that

$$\omega(C) = \omega(A) + \omega(B) \equiv \omega(A+B), \quad \forall \omega \qquad (1.3.10)$$

and (by the duality relation between states and observables) one may write $C = A + B \in \mathcal{O}$. For arbitrary pairs, A, B, the sum defined by the expectations (1.3.10) may not correspond to an element of $\mathcal{O}$ and therefore it leads to an extension of $\mathcal{O}$, (on which the states are positive linear functionals), for which the definition of the powers $(A+B)^n$ may not have a direct operational meaning. The possibility, adopted in the sequel, of introducing the powers of $A + B$ for *any* pair A, B, with the same algebraic properties of the powers of the generating observables and the extension of the states to them as *positive linear functionals* is therefore a non-trivial extrapolation over the strict physically motivated structure. The so-obtained extension of $\mathcal{O}$ will still be denoted by $\mathcal{O}$.

Clearly, from eqs. (1.3.5), (1.3.10), $||A + B||$ is well defined and

$$||A + B|| \le ||A|| + ||B||. \qquad (1.3.11)$$

Thus $|| \ ||$ is a norm on $\mathcal{O}$, which becomes a pre-Banach space. Technically it is convenient to consider the norm completion of $\mathcal{O}$, so that by a standard procedure one gets a real Banach space.

By definition of the norm, any state of the physical system satisfies

$$|\omega(A)| \le ||A||, \qquad (1.3.12)$$

i.e. any state is continuous with respect to the topology defined by the norm (*norm topology*) and therefore it has a unique continuous extension to the norm completion of $\mathcal{O}$, hereafter still denoted by $\mathcal{O}$.

The powers of the sum $A + B$ allow to define the following symmetric product

$$A \circ B \equiv \frac{1}{2}((A+B)^2 - A^2 - B^2) = B \circ A, \qquad (1.3.13)$$

which however is not guaranteed to be distributive and associative.

The so obtained structure is close to that advocated by Jordan, the so-called *Jordan Algebra*, [14] for which no topological structure is assumed, but

---

[14]P. Jordan, Zeit f. Phys. **80**, 285 (1933); L.J. Page, Jordan Algebras, in *Studies in Modern Algebra*, A.A. Albert ed., Prentice Hall 1963; N. Jacobson, *Structure and Representations of Jordan Algebras*, Am. Math. Soc. 1968; H. Upmeier, *Jordan Algebras in Analysis, Operator theory and Quantum Mechanics*, AMS 1980; H. Hanche-Olsen and E. Størmer, *Jordan Operator Algebras*, Pitman 1984.
A Jordan algebra is said to be special if the symmetric product arises from an associative product: $A \circ B = \frac{1}{2}(AB + BA)$: otherwise it is said to be exceptional. The analysis of exceptional Jordan algebras does not seem to have led anywhere for possible physical applications (for a general review see A.S. Wightman, Hilbert Sixth Problem: Mathematical Treatment of the Axioms of Physics, in *Proceedings of Symposia in Pure Mathematics*, Vol. 28, Am. Math. Soc. 1976).

the symmetric product is assumed to be distributive and weakly associative, i.e.

$$A^2 \circ (B \circ A) = (A^2 \circ B) \circ A.$$

The structure discussed above is rather close to that advocated by Segal for the description of quantum systems (*Segal system*), [15] for which additional continuity properties of the powers are assumed:

i) The square is continuous in the norm topology, i.e. $A_n \to A$ implies $A_n^2 \to A^2$,

ii) $||A^2 - B^2|| \leq \max(||A||^2, ||B||^2)$.

The Segal structure is recovered from the one discussed above under the (mild looking) assumption that the symmetric product is homogeneous, i.e.

$$A \circ (\lambda B) = \lambda(A \circ B), \quad \lambda \in \mathbf{R} \tag{1.3.14}$$

and then, by symmetry, also $(\lambda A) \circ B = \lambda(A \circ B)$. Such a property is certainly satisfied when A and B are polynomial functions of the same observable C, since then a (distributive and associative) product is defined and $A \circ B = \frac{1}{2}(AB + BA)$; the extension to the general case looks like a reasonable assumption.[16]

Now, homogeneity of the above product implies distributivity, (a property which is not assumed by Segal). In fact, eq. (1.3.13) gives

$$(A + B)^2 = A^2 + B^2 + 2A \circ B, \tag{1.3.15}$$

$$(A - B)^2 = A^2 + B^2 + 2A \circ (-B) = A^2 + B^2 - 2A \circ B$$

and therefore $A \circ B = \frac{1}{4}((A + B)^2 - (A - B)^2)$, as in Segal, and

$$A^2 + B^2 = \frac{1}{2}((A + B)^2 + (A - B)^2). \tag{1.3.16}$$

Then, by eq. (1.3.15)

$$2(A + B) \circ C - 2A \circ C - 2B \circ C$$

$$= [(A + B + C)^2 + A^2] + [B^2 + C^2] - [(A + B)^2 + (A + C)^2] - (B + C)^2,$$

and eq. (1.3.16) applied to the three sums of squares in square brackets gives the vanishing of the r.h.s.

Distributivity of the symmetric product and positivity of the states imply

$$0 \leq \omega\left((A + \lambda B)^2\right) = \omega\left((A + \lambda B) \circ (A + \lambda B)\right)$$

$$= \omega\left(A^2\right) + \lambda^2 \omega\left(B^2\right) + 2\lambda\omega\left(A \circ B\right), \quad \forall \lambda \in \mathbf{R},$$

[15] I. Segal, Ann. Math.(2), **48**, 930 (1947).

[16] G.G. Emch, *Algebraic Methods in Statistical Mechanics and Quantum Field Theory*, Wiley-Interscience 1972, pp. 44-47.

so that

$$|\omega\,(A \circ B)| \leq \omega\,(A^2)^{1/2}\,\omega\,(B^2)^{1/2} \tag{1.3.17}$$

and

$$||A \circ B|| \leq ||A||\,||B||. \tag{1.3.18}$$

Now, the continuity properties i) assumed by Segal follow easily. In fact, from distributivity one has $A_n^2 - A^2 = (A_n + A) \circ (A_n - A)$ and therefore

$$||A_n^2 - A^2|| \leq ||A_n - A||\,(||A_n - A|| + ||2A||),$$

which implies i). Equation ii) follows from the positivity of the squares and of the states, from the definition of the norm and from eq. (1.3.7), since $\forall a, b \in \mathbf{R}$, $|a^2 - b^2| \leq \max\,(a^2, b^2)$.

As shown by Segal, the above structure allows to recover most of the mathematical basis for the description of quantum systems, like the concept of compatible observables, the joint probability distribution for compatible observables etc. However, the mathematical language becomes easier if one makes the technical assumption that the so obtained Segal system $\mathcal{O}$ can be embedded in a complex extension $\mathcal{A}$ generated by complex linear combinations of elements of $\mathcal{O}$, such that
1) the symmetric product arises from an associative (but not necessarily commutative) product in $\mathcal{A}$, i.e. $\forall A, B \in \mathcal{O}$

$$A \circ B = \frac{1}{2}\,(AB + BA),$$

2) a * operation is defined on $\mathcal{A}$ with the properties that $\forall A, B \in \mathcal{O}$, $\lambda, \mu \in \mathbf{C}$

$$(\lambda A + \mu B)^* = \bar{\lambda}A + \bar{\mu}B,$$

$$(AB)^* = BA,$$

3) $\forall A \in \mathcal{A}$, A^*A is positive and the states can be extended from $\mathcal{O}$ to $\mathcal{A}$ by linearity as linear functionals, with the natural extension of positivity

$$\omega\,(A^*A) \geq 0, \quad \forall A \in \mathcal{A}\,,$$

and with the properties

$$||AB|| = \sup_\omega |\omega(AB)| \leq ||A||\,||B||, \quad ||A^*A|| = ||A^*||\,||A||. \tag{1.3.19}$$

Positivity, $\omega\,((\lambda A + \mathbf{1})^*(\lambda A + \mathbf{1})) \geq 0$, implies

$$\omega\,(A^*) = \overline{\omega(A)}\,, \quad ||A^*|| = ||A||, \quad \forall A \in \mathcal{A}\,. \tag{1.3.20}$$

The so-obtained extension $\mathcal{A}$ has the properties of a C^*-algebra with identity $\mathbf{1}$, $\mathcal{O}$ is the subset of *-invariant (also called *self-adjoint*) elements and $\mathcal{A}$ is generated by $\mathcal{O}$.[17]

Necessary and sufficient conditions have been given for the existence of such an extension, [18] but their physical interpretation is not transparent. Honestly, this should be regarded as a technically motivated assumption. A Segal system which allows such an extension is called special and exceptional otherwise; so far no one seems to have succeeded in giving an interesting physical application of exceptional Segal systems, which are actually very difficult to construct. In the following discussion, we shall therefore assume that the observables generate a special Segal system.

The arguments discussed in this section do not pretend to prove as a mathematical theorem that the general physical requirements on the set of observables necessarily lead to a C^*-algebraic structure, but they should provide sufficient motivations in favor of it. In any case, the above mathematical structure is by far more general than the concrete structure discussed in Sect. 2 for classical systems. As we shall see, the mathematical setting of quantum mechanics can be derived with a very strict logic solely from the C^*-algebraic structure of the observables and the *operational information of non-commutativity codified by the Heisenberg uncertainty relations* (Sect. 2.1). In this way one has a (in our opinion better motivated) alternative to the Dirac-Von Neumann axiomatic setting, which can actually be derived through the GNS theorem 2.2.4, the Gelfand-Naimark theorem 2.3.1 and Von Neumann theorem 3.2.2. For these reasons we adopt the following mathematical framework:

1. A physical system is *defined* by its C^*-*algebra* $\mathcal{A}$ *of observables* (with identity).

2. The states of the given physical system are identified by the measurements of the observables, i.e. a *state* is a *normalized positive linear functional on* $\mathcal{A}$. The set $\mathcal{S}$ of physical states separates the observables, technically one says that $\mathcal{S}$ is *full*, and conversely the observables separate the states.

In the mathematical literature, given a C^*-algebra $\mathcal{A}$, any normalized positive linear functional on it is by definition a state; here we allow the

---

[17]For an introduction to C^*-algebras see e.g. M. Takesaki, *Theory of Operator Algebras*, Vol. I, Springer 1979, Chap. I; R.V. Kadison and J.R. Ringrose, *Fundamentals of the Theory of Operator Algebras*, Vol. I, Academic Press 1983, Chap. 4; O. Bratteli and D.W. Robinson, *Operator Algebras and Quantum Statistical Mechanics*, Vol. I, Springer 1987, Sects. 2.1- 2.3; the basic textbook is J. Dixmier, C^*-*algebras* , North-Holland 1977. For the convenience of the reader a few basic notions about C^*-algebras are presented in the Appendices.

[18]D. Lowdenslager, Proc. Amer. Math. Soc. **8**, 88 (1957). For physically motivated conditions see E.M. Alfsen and F.W. Schultz, *Geometry of State Spaces of Operator Algbras*, Birkhäuser 2003.

possibility that the set $\mathcal{S}$ of states with physical interpretation (briefly called physical states) is full but smaller than the set of all the normalized positive linear functionals on $\mathcal{A}$.

Quite generally, given an abstract C^*-algebra $\mathcal{A}$, with identity $\mathbf{1}$, a positive linear functional ω on $\mathcal{A}$ is necessarily continuous with respect to the topology of the preassigned norm which makes $\mathcal{A}$ a C^*-algebra (see Appendix C, Proposition 1.6.3):

$$|\omega(A)| \leq ||A||\, \omega(\mathbf{1}).$$

Also, $\omega(A) \geq 0, \forall \omega$ implies $A = B^*B$ (Proposition 1.6.2). In the above presentation, these properties were obtained on the basis of the operational definition of states and observables.

The spectrum $\sigma(A)$ of an element $A \in \mathcal{A}$ is the set of all λ such that $\lambda \mathbf{1} - A$ does not have a two-sided inverse in $\mathcal{A}$. This is the purely algebraic version of the standard definition of spectrum for operators in a Hilbert space. An element A is said to be *normal* if it commutes with its adjoint A^*. If A is normal and $\lambda \in \sigma(A)$, then there exists at least one positive linear functional ω such that $\omega(A) = \lambda$ (see Appendix C). Thus the spectrum of a normal element A is a set of possible expectations of A. This implies that the set of all positive linear functionals on $\mathcal{A}$ separate the normal elements of $\mathcal{A}$ and therefore all the elements of $\mathcal{A}$, since any A can be written as a complex linear combination of normal elements $A = (A + A^*)/2 - i(iA - iA^*)/2$.

1.4 Appendix A : C^*-algebras

For the convenience of the reader, in this and in the following Appendices we recall a few basic notions about C^*-algebras.

A C^*-**algebra** $\mathcal{A}$ is
i) a linear associative algebra over the field $\mathbf{C}$ of complex numbers, i.e. a vector space over $\mathbf{C}$ with an associative product linear in both factors,
ii) a normed space, i.e. a norm $||\ ||$ is defined on $\mathcal{A}$:

$$||A|| \geq 0, \quad ||A|| = 0 \quad \Leftrightarrow \quad A = 0, \quad \forall A \in \mathcal{A} \ ,$$

$$||\lambda A|| = |\lambda|\,||A||, \quad \forall \lambda \in \mathbf{C},$$

$$||A + B|| \leq ||A|| + ||B||, \quad \forall A, B \in \mathcal{A} \ ,$$

with respect to which the product is continuous:

$$||AB|| \leq ||A||\,||B||, \tag{1.4.1}$$

and $\mathcal{A}$ is a complete space with respect to the topology defined by the norm (thus $\mathcal{A}$ is a Banach algebra),
iii) a *-(Banach) algebra, i.e. there is an involution $* : \mathcal{A} \to \mathcal{A}$,

$$(A + B)^* = A^* + B^*, \quad (\lambda A)^* = \bar{\lambda} A^*, \quad (AB)^* = B^* A^*, \quad (A^*)^* = A,$$

iv) with the property (C^*-*condition*)

$$||A^* A|| = ||A||^2. \tag{1.4.2}$$

The C^*-condition implies that

$$||A^*|| = ||A||. \tag{1.4.3}$$

In fact,
$$||A||^2 = ||A^* A|| \leq ||A^*||\,||A||,$$

i.e. $||A|| \leq ||A^*||$; on the other side, since $A = (A^*)^*$, by the same argument $||A^*|| \leq ||A||$.

An element $A \in \mathcal{A}$ is said to be *normal* if it commutes with its adjoint A^*; in this case the C^*-condition implies

$$||A^2|| = ||A||^2. \tag{1.4.4}$$

In fact,

$$||A^2||^2 = ||(A^*)^2 A^2|| = ||(A^* A)^* A^* A|| = ||A^* A||^2 = ||A||^4,$$

where in the second equality the normality of A has been used.

The above equation further implies that for a normal element A

$$||A^m|| = ||A||^m, \quad \forall m \in \mathbf{N}. \tag{1.4.5}$$

In fact, by iteration of eq. (1.4.4) one gets

$$||A^{2^k}|| = ||A||^{2^k}, \quad \forall k \in \mathbf{N}$$

and given $m \in \mathbf{N}$, one can find an n such that $m + n = 2^k$, so that

$$||A||^m \, ||A||^n = ||A||^{m+n} = ||A^{m+n}|| \leq ||A^m|| \, ||A^n|| \leq ||A||^m \, ||A||^n.$$

This means that the inequalities are actually equalities and the above equation is proved.

Given an element A of an algebra $\mathcal{A}$ with identity $\mathbf{1}$, as always assumed, its *spectrum* $\sigma(A)$ is defined as the set of all complex numbers λ such that $A - \lambda\mathbf{1}$ does not have a (two sided) inverse in $\mathcal{A}$.

Proposition 1.4.1 (Spectral radius formula) *Let $\mathcal{A}$ be a Banach algebra and $A \in \mathcal{A}$, then $\sigma(A)$ is a compact not empty set and*

$$\sup_{\lambda \in \sigma(A)} |\lambda| = \lim_{n \to \infty} ||A^n||^{1/n} \leq ||A||. \tag{1.4.6}$$

If $\mathcal{A}$ is a C^*-algebra and A is normal, then the above inequality becomes an equality.

Proof. We start by showing that the limit on the r.h.s. exists. For this purpose, let

$$r \equiv \inf_{n \in \mathbf{N}} ||A^n||^{1/n}.$$

Clearly, $r \leq ||A^n||^{1/n} \leq ||A||$, $\forall n \in \mathbf{N}$, and therefore

$$r \leq \lim_{n \to \infty} \inf ||A^n||^{1/n}.$$

Now, let $\varepsilon > 0$ and choose m such that $||A^m||^{1/m} < r + \varepsilon$. For any $n \in \mathbf{N}$, $\exists k_n \in \mathbf{N}$ such that $n = k_n m + l_n$, $l_n \in \mathbf{N}$, $0 \leq l_n \leq m$; then

$$||A^n||^{1/n} = ||A^{k_n m} A^{l_n}||^{1/n} \leq ||A^m||^{k_n/n}||A||^{l_n/n} \leq (r + \varepsilon)^{mk_n/n}||A||^{l_n/n}.$$

By construction $\lim_{n \to \infty} mk_n/n = 1$, $\lim_{n \to \infty} l_n/n = 0$, so that

$$\lim_{n \to \infty} \sup ||A^n||^{1/n} \leq r + \varepsilon.$$

Since ε was arbitrary, one has

$$\lim_{n \to \infty} \sup ||A^n||^{1/n} = \lim_{n \to \infty} \inf ||A^n||^{1/n},$$

i.e. the limit exists. The equality for normal elements of a C^*-algebra follows trivially from eq. (1.4.5).

To conclude the proof, we note that the existence of a norm allows an extension of the standard analytic calculus to Banach algebras (see e.g. the quoted book by Gelfand, Raikov and Shilov); in particular, by the extension of the Cauchy-Hadamard theorem of elementary analytic calculus, r^{-1} is the radius of convergence of the series $\mathbf{1} + zA + z^2A^2 + ..., z \in \mathbf{C}, A \in \mathcal{A}$, which converges to $(\mathbf{1} - zA)^{-1}$ for $|z| < r^{-1}$ and has a singularity for $|z| = r^{-1}$. Thus, $(\mu\mathbf{1} - A)^{-1}$ exists if $|\mu| > r$, and $r = \sup_{\lambda \in \sigma(A)} |\lambda|$; moreover $\sigma(A)$ is closed because the analyticity domain $\mathbf{C}/\sigma(A)$ of $(A - z\mathbf{1})^{-1}$ is open.

By a similar argument, $\sigma(A)$ cannot be empty, otherwise $(\lambda\mathbf{1} - A)^{-1}$ would be an entire function in the whole complex λ plane, vanishing for $|\lambda| \to \infty$ and therefore zero everywhere, contrary to the existence of A^{-1} (implied by $\sigma(A) = \emptyset$).

The above Proposition implies that if all elements, except 0, of a Banach algebra $\mathcal{A}$ are invertible, then $\mathcal{A}$ is isomorphic to the complex numbers; in fact, if $\lambda \in \sigma(A) \neq \emptyset$, then $\lambda\mathbf{1} - A$ is not invertible and therefore it must be 0, i.e. $A = \lambda\mathbf{1}$ (*Gelfand-Mazur theorem*).

A family $\mathcal{F} = \{A_\alpha, \alpha \in I\}$ is said to *generate* a (normed) algebra $\mathcal{A}$ if the polynomials of $\mathcal{F}$ are dense in $\mathcal{A}$.

1.5   Appendix B: Abelian C^*-algebras

A C^*-algebra $\mathcal{A}$ is called *abelian* or *commutative* if the product is commutative.

Definition 1.5.1 *A* **multiplicative linear functional** *m on a commutative Banach algebra $\mathcal{A}$ is a homomorphism of $\mathcal{A}$ into $\mathbf{C}$, i.e. a mapping which preserves all the algebraic properties:*

$$m(AB) = m(A)m(B), \quad m(A + B) = m(A) + m(B). \tag{1.5.1}$$

Clearly $m(\mathbf{1}) = 1$, if $m \not\equiv 0$.

Definition 1.5.2 *A linear subspace I of an algebra $\mathcal{A}$ is called a* **left** *(respectively* **right***) **ideal** *if it is stable under left (resp. right) multiplication by elements of $\mathcal{A}$. I is* **proper** *if it is properly contained in $\mathcal{A}$ ($I \neq \mathcal{A}$), and it is* **maximal** *if it is not properly contained in a proper ideal.*

For commutative algebras left and right ideals coincide and are simply called ideals. Clearly, if $\mathcal{A}$ has an identity $\mathbf{1}$, as always assumed, an ideal

I is proper iff $\mathbf{1} \notin I$, $(\mathbf{1} \in I \Rightarrow \mathcal{A} \mathbf{1} \subseteq I$, i.e. $I = \mathcal{A})$. Hence, for Banach algebras, the closure of a proper ideal I is a proper ideal (if $\mathbf{1} \in \bar{I}$, $\exists x \in I$, with $\|\mathbf{1} - x\| < 1$, $x = \mathbf{1} - (\mathbf{1} - x)$ is invertible and $x^{-1} x = \mathbf{1} \in I$) and therefore maximal ideals are closed and each proper ideal is contained in a proper maximal ideal.

Proposition 1.5.3 *For a commutative Banach algebra $\mathcal{A}$ there is a one to one correspondence between the set $\Sigma(\mathcal{A})$ of multiplicative linear functionals and proper maximal ideals of $\mathcal{A}$.*

Furthermore, given $A \in \mathcal{A}$, $\lambda \in \sigma(A)$ iff there exists a multiplicative linear functional $m \in \Sigma(\mathcal{A})$ such that $m(A) = \lambda$.

Proof. Each $m \in \Sigma(\mathcal{A})$ defines a proper ideal $K \equiv \ker(m)$. Since $\forall [A], [B] \in \mathcal{A}/K$, $m([A]) = m([B])$ implies $[A] = [B]$, it follows that $\mathcal{A}/K$ is isomorphic to $\mathbf{C}$; it is therefore a field and then it cannot contain any proper ideal, since an invertible element A of a commutative Banach algebra cannot belong to any proper ideal I (otherwise $\mathbf{1} = AA^{-1} \in I$). This excludes the existence of a maximal proper ideal I properly containing K, because otherwise I/K would be a proper ideal of $\mathcal{A}/K$; thus K is maximal.

Conversely, given a maximal proper ideal K of $\mathcal{A}$, $\mathcal{A}/K$ is a Banach space (with $\|[A]\| = \inf_{k \in K} \|A + k\|$), since K is closed, and actually a Banach algebra, since K is an ideal. Furthermore, since K is maximal, $\mathcal{A}/K$ cannot contain a proper ideal, otherwise its inverse image in $\mathcal{A}$ would be a proper ideal which properly contains K. Thus, all elements of $\mathcal{A}/K$, except 0, are invertible, since a non-invertible element A of a commutative Banach algebra $\mathcal{A}$ belongs to the ideal $A\mathcal{A}$, which does not contain $\mathbf{1}$ and therefore is proper. Hence, by the Gelfand-Mazur theorem, $\mathcal{A}/K$ is isomorphic to $\mathbf{C}$, and the homomorphism $m : \mathcal{A} \rightarrow \mathcal{A}/K \rightarrow \mathbf{C}$ defines a unique multiplicative linear functional with $\ker(m) = K$.

For the second part of the Proposition, if $\lambda \in \sigma(A)$ then $\lambda\mathbf{1} - A$ is not invertible and therefore it belongs to the proper ideal $I \equiv (\lambda\mathbf{1} - A)\mathcal{A}$. Let J be a maximal ideal containing I and m_J the corresponding multiplicative linear functional, then $\ker(m_J) = J \supseteq I$ so that $m_J(\lambda\mathbf{1} - A) = 0$, i.e. $m_J(A) = \lambda$.

Conversely, if $\exists m$, with $m(A) = \lambda$, then $\lambda\mathbf{1} - A$ is not invertible i.e. $\lambda \in \sigma(A)$, since otherwise

$$1 = m(\mathbf{1}) = m((\lambda\mathbf{1} - A)(\lambda\mathbf{1} - A)^{-1}) = m(\lambda\mathbf{1} - A)\, m((\lambda\mathbf{1} - A)^{-1}) = 0.$$

Because of the above relation between $\Sigma(\mathcal{A})$ and the points of the spectra of the elements of $\mathcal{A}$, $\Sigma(\mathcal{A})$ is called *the (Gelfand) spectrum of $\mathcal{A}$*. Indeed, if $\mathcal{A}$ is generated by a single element A, i.e. the linear span of the powers of A is dense in $\mathcal{A}$, then $\Sigma(\mathcal{A}) = \sigma(A)$; in fact, the above Proposition, establishes a correspondence between $\sigma(A)$ and $\Sigma(\mathcal{A})$, which is actually one to one since if $m_1(A) = m_2(A)$, then m_1 and m_2 coincide on the polynomials of A and, since the latter are dense, on the whole of $\mathcal{A}$.

By the spectral radius formula, the above relation between $\Sigma(\mathcal{A})$ and $\sigma(A)$ implies that multiplicative linear functionals are continuous:

$$|m(A)| \leq \sup_{\lambda \in \sigma(A)} |\lambda| \leq ||A||,$$

with an equality on the right if $\mathcal{A}$ is a (abelian) C^*-algebra.

Proposition 1.5.4 *Let $\mathcal{A}$ be a C^*-algebra (with identity), then any bounded linear functional m on $\mathcal{A}$, with $m(\mathbf{1}) = 1 = ||m||$, satisfies*

$$m(A^*) = \overline{m(A)}. \tag{1.5.2}$$

Proof. First we prove that if $A = A^*$, then $m(A)$ is real. Indeed, putting $m(A) = a + i\,b$, $a, b \in \mathbf{R}$, we have $\forall c \in \mathbf{R}$

$$b^2 + c^2 + 2bc = |b + c|^2 \leq |a + i(b + c)|^2 = |m(A + ic\mathbf{1})|^2$$

$$\leq ||A + ic\mathbf{1}||^2 = ||A^*A + c^2|| \leq ||A||^2 + c^2,$$

where the C^* condition has been used. The above inequality requires $b = 0$. Now, a generic $A \in \mathcal{A}$ can be written as a linear combination of self-adjoint elements: $A = (A + A^*)/2 - i(iA - iA^*)/2$, so that the result follows by linearity.

If $\mathcal{A}$ is a commutative C^*-algebra is generated by A and A^*, then by exploiting the above property of multiplicative linear functionals and the general argument for abelian Banach algebras, one has that $\Sigma(\mathcal{A}) = \sigma(A)$. More generally, if $\mathcal{A}$ is a commutative C^*-algebra generated by (algebraically independent) $A_1, A_2, ..., A_n, A^*_1, ..., A^*_n$, then

$$\Sigma(\mathcal{A}) = \times_i \sigma(A_i).$$

Theorem 1.5.5 *(Gelfand-Naimark characterization of abelian C^*-algebras) An abelian C^*-algebra $\mathcal{A}$ (with identity) is isometrically isomorphic to the C^*-algebra of continuous functions on a compact Hausdorff topological space, which is the Gelfand spectrum of $\mathcal{A}$ with the topology induced by the weak * topology.*

Proof. By duality, each $A \in \mathcal{A}$ defines a function $\tilde{A}$ on $\Sigma(\mathcal{A})$, called the Gelfand transform of A, by $\tilde{A}(m) \equiv m(A)$ and clearly

$$(\tilde{A} + \tilde{B})(m) = \tilde{A}(m) + \tilde{B}(m), \quad \mu\tilde{A}(m) = \widetilde{\mu A}(m), \ \forall \mu \in \mathbf{C}, \tag{1.5.3}$$

$$(\tilde{A}\tilde{B})(m) = \widetilde{AB}(m) = \tilde{A}(m)\,\tilde{B}(m), \quad \overline{\tilde{A}(m)} = (\tilde{A^*})(m) \equiv (\tilde{A})^*(m). \tag{1.5.4}$$

Thus, the functions $\tilde{A}$, for $A \in \mathcal{A}$, form an abelian *-algebra $\tilde{\mathcal{A}}$. By the above Proposition, for each $m \in \Sigma(\mathcal{A})$, $m(A)$ is a point of $\sigma(A)$, which is a closed set, and

$$|\tilde{A}(m)| = |m(A)| \leq \sup_{\lambda \in \sigma(A)} |\lambda| = ||A||,$$

$$||\tilde{A}||_\infty \equiv \sup_{m \in \Sigma(\mathcal{A})} |\tilde{A}(m)| = \sup_{\lambda \in \sigma(A)} |\lambda| = ||A||. \qquad (1.5.5)$$

Thus $\mathcal{A}$ is isometrically isomorphic to $\tilde{\mathcal{A}}$.

We shall now show that $\Sigma(\mathcal{A})$ is a compact topological space and that $\tilde{\mathcal{A}} = C(\Sigma(\mathcal{A}))$. For this purpose, we note that $\Sigma(\mathcal{A})$ is a closed subset of the closed unit ball $\mathcal{B}$ of the set $\mathcal{A}^*$ of continuous linear functionals on $\mathcal{A}$. Indeed
i) each $m \in \Sigma(\mathcal{A})$ is a continuous functional since $|m(A)| \leq ||A||$, and, since $1 \in \mathcal{A}$ and $m(1) = 1$,

$$||m||_\infty = \sup_{A \in \mathcal{A}} |m(A)|/||A|| = 1,$$

i.e. $m \in \mathcal{B}$.
ii) the weak * topology on $\mathcal{A}^*$ is defined by the following basic neighborhoods: given $\varepsilon > 0, A_1, ..., A_n \in \mathcal{A}$, the neighborhood of $\bar{l}$ is

$$U_{A_1,...,A_n}(\bar{l}; \varepsilon) = \{l \in \mathcal{A}^* : |l(A_i) - \bar{l}(A_i)| < \varepsilon, \quad i = 1, ..., n\}.$$

With respect to such a topology $\mathcal{A}^*$ is a Hausdorff topological space and the unit ball $\mathcal{B} \subset \mathcal{A}^*$ is compact, by the Alaoglu-Banach theorem[19]. The topology induced on $\Sigma(\mathcal{A})$ by the weak * topology is called the Gelfand topology and it is the weakest topology under which all the functions $\tilde{A}(m)$ are continuous. Clearly, $\Sigma(\mathcal{A})$ is a Hausdorff space because the weak * topology is Hausdorff.
iii) it remains to prove that $\Sigma(\mathcal{A})$ is a weak * closed set of $\mathcal{B}$. As a matter of fact, if $m_\alpha \in \Sigma(\mathcal{A})$ and $m_\alpha \to l$ in $\mathcal{B}$, then $\forall A, B \in \mathcal{A}$

$$m_\alpha(AB) = m_\alpha(A) m_\alpha(B) \quad \Rightarrow \quad l(AB) = l(A) l(B),$$

i.e. l is multiplicative.

Finally, since by definition $\tilde{\mathcal{A}}$ separates the points of $\Sigma(\mathcal{A})$, and it is closed by eq. (1.5.5), by the Stone-Weierstrass theorem it is the whole $C(\Sigma(\mathcal{A}))$.

Examples. To better grasp the properties of C^*-algebras it is instructive to work out the following Exercises.

1. Let X be a compact Hausdorff topological space and $C(X)$ the C^*-algebra of the continuous functions on X.

---

[19]See e.g. N. Dunford and J.T. Schwartz, *Linear Operators. Part I: General Theory*, Interscience 1958; M. Reed and B. Simon, *Methods of Modern Mathematical Physics*, Academic Press, Vol. 1, p. 115.

a. Determine the spectrum $\sigma(f)$ for $f \in C(X)$ and show that $\forall F \in C(\mathbf{C})$, $\sigma(F(f)) = F(\sigma(f))$.
b. Verify explicitly the validity of eq. (1.4.6).
c. Determine independently the proper maximal ideals of $C(X)$ and its multiplicative linear functionals; verify Proposition 1.5.3 and that $\Sigma(C(X))$ $= X$. [Hints: If, given a proper maximal ideal I, $\forall x \in X$, there is a $f_x \in I$ such that $f_x(x) \neq 0$, then, by exploiting the compactness of X, one could construct a never vanishing $h \in I$ and $h\,h^{-1} = 1 \in I$, so that I is not proper. If the support K of a multiplicative measure μ on X (i.e. the smallest compact set such that $\mu(f) = 0$, if supp $f \cap K = \emptyset$) contains two disjoint open sets K_1, K_2, then $\exists g \in C(X)$, supp $g \subseteq K_2$, such that $\mu(g) \neq 0$ and $\forall f \in C(X)$, supp $f \subseteq K_1$, $0 = \mu(fg) = \mu(f)\mu(g)$, i.e. $\mu(f) = 0$.]
2. Let $\mathcal{M}$ be the set of diagonal $n \times n$ matrices. Verify that $\mathcal{M}$ is a C^*-algebra. Determine the spectrum of $M \in \mathcal{M}$ and the Gelfand spectrum of $\mathcal{M}$.

1.6 Appendix C: Spectra and states

We discuss general properties of the states of a C^*-algebra $\mathcal{A}$ and their relation with the spectra of the normal elements of $\mathcal{A}$.

Proposition 1.6.1 *Let $\mathcal{A}, \mathcal{B}$ be C^*-algebras and $\mathcal{A} \subset \mathcal{B}$, then for any $A \in \mathcal{A}$, the set $\sigma_{\mathcal{A}}(A)$ of $\lambda \in \mathbf{C}$ such that $\lambda 1 - A$ is not invertible in $\mathcal{A}$ coincides with*

$$\sigma(A) = \{\lambda \in \mathbf{C} : \lambda 1 - A \text{ is not invertible in } \mathcal{B}\}.$$

Proof. In fact, if $(\lambda 1 - A)^{-1}$ exists in $\mathcal{B}$, it can be expressed as a convergent power series, i.e. it is the norm limit of partial sums each belonging to $\mathcal{A}$, so that also $(\lambda 1 - A)^{-1}$ belongs to $\mathcal{A}$.

Proposition 1.6.2 *Given a normal element A of a C^*-algebra $\mathcal{A}$, then any continuous function $F = F(A, A^*)$ of A, A^* defines an element of $\mathcal{A}$ and $\sigma(F) = \tilde{F}(\sigma(A))$, where $\tilde{\ }$ denotes the Gelfand transform.*

Proof. Let $\mathcal{A}_A$ be the abelian algebra generated by A, A^* and 1; by the Gelfand-Naimark theorem, F defines an element of $\mathcal{A}_A \subset \mathcal{A}$ and, by the preceding Proposition, $\sigma(F) = \sigma_{\mathcal{A}_A}(F) = \sigma(\tilde{F}) = \text{Range } \tilde{F} = \tilde{F}(\text{Range } \tilde{A}) = \tilde{F}(\sigma(A))$.

By Proposition 1.6.2 the functional calculus for normal A follows from the calculus on functions; e.g. $\sigma(A) \subseteq \mathbf{R}_+$ iff $\tilde{A} \geq 0$ and in this case $\tilde{A}^{1/2} \geq 0$ defines $A^{1/2}$. Similarly, the decomposition $\tilde{A} = \tilde{A}_+ - \tilde{A}_-$, with $\tilde{A}_\pm \geq 0$, $\tilde{A}_+ \tilde{A}_- = 0$ yields $A = A_+ - A_-$, $\sigma(A_\pm) \geq 0$, $A_+ A_- = 0$.

The following properties are equivalent and define the set $\mathcal{A}_+$ of the *positive* elements, [20]

$$1)\ \sigma(A) \subseteq [0, ||A||],\quad 2)\ A = B^2,\ B = B^*,\quad 3)\ ||\mathbf{1} - A/||A||\ || \le 1,$$

$$4)\ A = C^* C.$$

$\mathcal{A}_+$ is a closed convex cone, since 3) is stable under closure and multiplication of A by positive numbers, and $\forall A$, B, with $||A|| = ||B|| = 1$, one has $||\mathbf{1} - (A+B)/2|| \le \frac{1}{2}(||\mathbf{1} - A|| + ||\mathbf{1} - B||) \le 1$.

Proposition 1.6.3 *A linear functional ω on a C^*-algebra $\mathcal{A}$, with identity, is positive iff: 1) ω is bounded, 2) $||\omega|| = \omega(\mathbf{1})$.*

Proof. Let ω be positive, then by the Cauchy-Schwarz' inequality

$$|\omega(A^*B)|^2 \le \omega(A^*A)\,\omega(B^*B),$$

which implies

$$|\omega(A)\ |^2 \le \omega(\mathbf{1})\,\omega(A^*A).$$

Thus, to get continuity it suffices to prove that

$$\omega(A^*A) \le ||A||^2\,\omega(\mathbf{1}),$$

i.e. that for any positive B, $\omega(B) \le ||B||\,\omega(\mathbf{1})$. This follows easily from the discussion after Proposition 1.6.2, in particular from the equivalent definitions of positive elements. In fact, one has that

$$||B||\mathbf{1} - B \ge 0$$

and therefore, by the positivity of ω,

$$\omega(B) \le ||B||\,\omega(\mathbf{1}).$$

Conversely, let ω be bounded and $||\omega|| = \omega(\mathbf{1})$; it suffices to take $\omega(\mathbf{1}) = 1$ and consider A with $||A|| = 1$. By Proposition (1.5.4) $\omega(A^*A)$ is real and one has

$$|1 - \omega(A^*A)| = |\omega(\mathbf{1} - A^*A)| \le ||\mathbf{1} - A^*A|| \le 1,$$

where the first inequality follows from the continuity of ω and the last from the positivity of A^*A (see the equivalent property 3 above). This requires $\omega(A^*A) \ge 0$.

---

[20]Clearly, 1) $\Leftrightarrow$ 2), by the existence of $A^{1/2}$. 1) $\Rightarrow$ 3), since $\sigma(\mathbf{1} - A/||A||) = 1 - \sigma(A)/||A|| \subseteq [0, 1]$ and conversely, $||\mathbf{1} - A/||A||\ || \le 1$ implies $\sigma(\mathbf{1} - A/||A||) \subseteq [-1, 1]$, i.e. $\sigma(A) \subseteq [0, 2||A||]$. For the equivalence 4) $\Leftrightarrow$ 1), one notes that $A = a_+^2 - a_-^2$, $a_\pm = A_\pm^{1/2}$ and $(Ca_-)^*(Ca_-) = a_-(a_+^2 - a_-^2)a_- = -a_-^4 \in -\mathcal{A}_+$. On the other hand, by writing $D \equiv Ca_- = a_1 + ia_2$, $a_i = a_i^*$, one has $DD^* = -D^*D + 2a_1^2 + 2a_2^2 \in \mathcal{A}_+$. Now, if $\lambda \notin \sigma(D^*D) \cup \{0\}$, $\exists E = (D^*D - \lambda\mathbf{1})^{-1}$ and the identities $(DD^* - \lambda\mathbf{1})(DED^* - 1) = \lambda = (DED^* - 1)(DD^* - \lambda\mathbf{1})$ imply $\lambda \notin \sigma(DD^*) \cup \{0\}$ and by symmetry $\sigma(D^*D) \cup \{0\} = \sigma(DD^*) \cup \{0\}$; then, one gets a contradiction, unless $a_- = 0$.

Proposition 1.6.4 *Let A be a normal element of a C^*-algebra $\mathcal{A}$, then $\lambda \in \sigma(A)$ iff there exists a positive linear functional ω on $\mathcal{A}$ such that, for any polynomial $\mathcal{P}(A, A^*)$, one has $\omega\left(\mathcal{P}(A, A^*)\right) = \mathcal{P}(\lambda, \bar{\lambda})$.*

Proof. Let $\mathcal{A}_A$ be the abelian algebra generated by A, A^* and the identity. By Proposition 1.6.1, $\lambda \in \sigma_{\mathcal{A}}(A)$ and by Proposition 1.5.3 there exists a multiplicative linear functional ω_A on $\mathcal{A}_A$ (and therefore $\omega_A(\mathbf{1}) = 1$), such that $\omega_A(A) = \lambda$, $\omega_A(A^*) = \bar{\lambda}$ and therefore $\omega_A(\mathcal{P}(A, A^*)) = \mathcal{P}(\lambda, \bar{\lambda})$. Furthermore, ω_A is positive on $\mathcal{A}_A$ and therefore, by Proposition 1.6.3, is a bounded linear functional with $||\omega_A|| = \omega_A(\mathbf{1}) = 1$. Since $\mathcal{A}_A$ is a closed subalgebra of $\mathcal{A}$, by the Hahn-Banach theorem ω_A has an extension ω to $\mathcal{A}$ with $||\omega|| = ||\omega_A|| = \omega_A(\mathbf{1}) = 1 = \omega(\mathbf{1})$. By Proposition 1.6.3, ω is a positive linear functional on $\mathcal{A}$ and coincides with ω_A on $\mathcal{A}_A$.
Conversely, if $\omega\left(\mathcal{P}(A, A^*)\right) = \mathcal{P}(\lambda, \bar{\lambda})$, then ω is a multiplicative linear functional on $\mathcal{A}_A$ and $\omega(A) = \lambda$. Then, by Proposition (1.5.3), $\lambda \in \sigma(A)$.

As an immediate consequence of the above Proposition we have

Proposition 1.6.5 *The positive linear functionals on a C^*-algebra $\mathcal{A}$ separate the elements of $\mathcal{A}$.*

Proof. Let $A, B \in \mathcal{A}, A \neq B$. Then, $||A - B|| \neq 0$ and, putting $A - B = A_1 + iA_2$, with $A_i = A_i^*$, $i = 1, 2$, one has $||A_1|| + ||A_2|| > 0$. Without loss of generality we consider the first case $||A_1|| \neq 0$. Then $\lambda = ||A_1|| \in \sigma(A_1)$ and by the preceding Proposition there exists a state ω such that $0 \neq \lambda = \omega(A_1) = \operatorname{Re} \omega(A - B)$.

Proposition 1.6.6 *An element A of a C^*-algebra $\mathcal{A}$ is positive iff $\omega(A) \geq 0$, for all positive linear functionals ω.*

Proof. In fact $\omega(A) \geq 0$, $\forall \omega$, implies $\omega(A - A^*) = 0, \forall \omega$, i.e. $A = A^*$, and therefore if $\lambda \in \sigma(A)$, $\exists \omega$ such that $\omega(A) = \lambda$, (see the proof of Proposition 1.6.4 above) i.e. $\lambda \geq 0$ and $\sigma(A) \subseteq [0, ||A||]$.

Positivity allows to introduce a natural ordering of linear functionals: given two positive linear functional ω_1, ω_2, ω_1 is said to majorize ω_2, briefly $\omega_1 \geq \omega_2$, if $\omega_1 - \omega_2$ is a positive linear functional.

The functional calculus developed above allows to prove the spectral representation of a bounded self-adjoint operator A on a Hilbert space $\mathcal{H}$, (the so-called *spectral theorem*), which generalizes the standard representation of an $n \times n$ hermitian matrix M

$$M = \sum_{i=1}^{n} \lambda_i P_i, \tag{1.6.1}$$

where λ_i are the eigenvalues of M and P_i are the projections on the corresponding eigenvectors (if $\lambda_i = \lambda_j$, P_i and P_j have to be chosen as two independent projections on the corresponding two dimensional space).

Theorem 1.6.7 *For a bounded self-adjoint operator A one can write the following spectral representation*

$$A = \int_{\sigma(A)} \lambda dP(\lambda),$$

with $dP(\lambda)$ a projection valued measure defined on the spectrum of A.

A simple proof of the spectral theorem exploits the construction of projection operators $P(\Delta)$ which correspond to the characteristic functions of intervals $\Delta = [a, b) \subseteq \sigma(A)$ (see below for their explicit construction). They are characterized by the property of projecting on the subspace $\mathcal{H}_\Delta \subseteq \mathcal{H}$, such that the corresponding expectations of A are in Δ; thus

$$a\, P(\Delta) \leq A\, P(\Delta) \leq b\, P(\Delta). \tag{1.6.2}$$

Clearly, $AP(\Delta) = 0$ if $\Delta \cap \sigma(A) = \emptyset$, and if $\cup_i \Delta_i = \sigma(A)$, $\Delta_i \cap \Delta_j = \emptyset$, for $i \neq j$, then

$$P(\cup \Delta_i) = \sum P(\Delta_i) = \mathbf{1}. \tag{1.6.3}$$

Now, if $\nu_k \in [\lambda_k, \mu_k) \equiv \Delta_k$, $\delta \equiv \max |\mu_k - \lambda_k|$, and $\cup \Delta_k = \sigma(A)$ one has from eq. (1.6.2)

$$\sum_k (\lambda_k - \nu_k) P(\Delta_k) \leq A - \sum_k \nu_k P(\Delta_k) \leq \sum_k (\mu_k - \nu_k) P(\Delta_k)$$

and

$$\delta \leq A - \sum_k \nu_k P(\Delta_k) \leq \delta,$$

i.e. the Riemann sums $\sum \nu_k P(\Delta_k)$ converge to A as $\delta \to 0$. Thus, as in the ordinary case, one may introduce the operator valued integral

$$A = \lim_{\delta \to 0} \sum \nu_k P(\Delta_k) = \int_{\sigma(A)} \lambda\, dP(\lambda), \tag{1.6.4}$$

which gives the spectral representation of A in terms of the projection valued measure $dP(\lambda)$.

The existence of the projections $P(\Delta)$ can be argued by explicit construction. Since $A^2 \geq 0$, by the remark after Proposition 1.6.2, one may define the positive square root $\sqrt{A^2}$ and $A - \sqrt{A^2} \leq 0$. The projection P_+ on the subspace $\mathcal{H}_+ = \{x \in \mathcal{H};\, (A - \sqrt{A^2})x = 0\}$ has the meaning of the projection on the subspace on which the expectations of A are positive, i.e. $AP_+ \equiv A_+ \geq 0$. Clearly $\mathbf{1} - P_+$ projects on the subspace on which the expectations of A are negative. By the same reasoning applied to the operator $A_\lambda = A - \lambda \mathbf{1}$ one defines the projection $P_+(\lambda)$, (corresponding to the subspace on which $A \geq \lambda$), and finally for $\Delta = [\lambda, \mu)$, $P(\Delta) \equiv P_+(\lambda) - P_+(\mu)$ is the required projection.

A more elegant and compact proof of the spectral theorem can be obtained by exploiting the functional calculus and Proposition 1.6.2.

First, if A is a self-adjoint (more generally a normal) element of a C^*−algebra $\mathcal{A}$, a possible realization of the Gelfand isomorphism $\mathcal{A}_A \to C(\Sigma(\mathcal{A}_A)) = C(\sigma(A))$ is given by $\hat{A}(\lambda) = \lambda$, $\hat{F}(A) = F(\lambda)$, for any continuous function F. In fact, in this way one realizes an isometric isomorphism between the polynomial *-algebra generated by A and the *-algebra of polynomials on $C(\sigma(A))$ and by continuity such an isomorphism extends to the continuous functions on $C(\sigma(A))$.

The second step is to note that a vector $x \in \mathcal{H}$ defines a positive linear functional $(x, F(A) x)$ on the the algebra of continuous functions F of A, and therefore a positive linear functional on $C(\sigma(A))$. Then, by the Riesz-Markov theorem there exists a (unique) regular Borel measure μ_x such that

$$(x, F(A) x) = \int_{\sigma(A)} F(\lambda) \, d\mu_x(\lambda). \qquad (1.6.5)$$

By the polarization identity

$$(x, Ay) = \frac{1}{4}[(x + y, A(x + y)) - (x - y, A(x - y))$$

$$-i(x + iy, A(x + iy)) + i(x - iy, A(x - iy))] \qquad (1.6.6)$$

also $(x, F(A)y)$ is a continuous linear functional on $C(\sigma(A))$ and therefore expressible in terms of a complex measure μ_{xy} on $C(\sigma(A))$

$$(x, F(A)y) = \int_{\sigma(A)} F(\lambda) \, d\mu_{xy}(\lambda). \qquad (1.6.7)$$

Such a spectral representation allows to extend the Gelfand Naimark isomorphism $F(A) \to F(\lambda)$ from the continuous functions of A to the bounded Borel functions B, by putting

$$(x, B(A)y) \equiv \int_{\sigma(A)} B(\lambda) \, d\mu_{xy}(\lambda). \qquad (1.6.8)$$

In fact, the r.h.s. is a continuous sesquilinear form on $\mathcal{H}$ and by the Riesz lemma identifies a unique operator B.

In particular, if $\chi(\Delta)$ denotes the characteristic function of the Borel set Δ, the corresponding operator $P(\Delta)$ (called a *spectral projection*) satisfies

i) (*projection*) $P(\Delta) = P(\Delta)^*$, $\quad P(\Delta_k) P(\Delta_j) = P(\Delta_k \cap \Delta_j)$

ii) $P(\emptyset) = 0$, $\quad P(\sigma(A)) = 1$

iii) (σ-*additivity*) If $\Delta_j \cap \Delta_k = \emptyset$, $\forall j \neq k$

$$P(\cup \Delta_k) = s - \lim_{N \to \infty} \sum_{k=1}^{N} P(\Delta_k).$$

Such a system of projections is called a partition of unity, since if $\cup \Delta_k = \sigma(A)$, $\Delta_j \cap \Delta_k = \emptyset$, σ-additivity gives

$$1 = \sum_k P(\Delta_k).$$

Now, for any fixed $x \in \mathcal{H}$, the expectation $\mu_x(\Delta) \equiv (x, P(\Delta)x)$ defines a measure on the Borel sets of $\sigma(A)$ and by a standard measure theoretical argument

$$(x, F(A)x) = \int_{\sigma(A)} F(\lambda) \, d(x, P(\lambda)x), \qquad (1.6.9)$$

where $d(x, P(\lambda)x)$ denotes the integration with respect to μ_x. Riesz lemma then gives the spectral representation of $F(A)$ and, in particular,

$$A = \int_{\sigma(A)} \lambda \, dP(\lambda), \qquad (1.6.10)$$

where $dP(\lambda)$ is the projection valued measure defined by the family of projections $P(\Delta)$.

As a simple application we have obtain Stone's theorem

Theorem 1.6.8 *A bounded self-adjoint operator A defines a one-parameter group of strongly continuous unitary operators $U(t)$, $t \in \mathbf{R}$, of which A is the generator in the sense that*

$$\text{strong-}\lim_{t \to 0} t^{-1} \left(U(t) - \mathbf{1}\right) = i \, A. \qquad (1.6.11)$$

Proof. The unitary group $U(t)$ is defined by the spectral integral

$$U(t) = e^{i\,A\,t} = \int_{\sigma(A)} e^{i\,\lambda\,t} \, dP(\lambda), \qquad (1.6.12)$$

with $dP(\lambda)$ the spectral measure defined by A. The group law and unitarity follow from the Gelfand isomorphism and the spectral representation.

The strong continuity and eq. (1.6.11) follows from the dominated convergence theorem respectively applied the the sequence $f_t(\lambda) \equiv -1 + \exp(i\lambda t)$ which converges pointwise to zero, for $t \to 0$, and it is dominated by 2 and to the sequence $\Delta f_t \equiv t^{-1}(\exp(i\lambda t) - 1)$ which converges pointwise to $i\lambda$ and it is dominated by $|\lambda|$, for $t \to 0$.

Chapter 2

Mathematical description of a quantum system

2.1 Heisenberg uncertainty relations and non-abelianess

The puzzling situation, briefly discussed in Sect. 1.1, which characterizes the conflict between classical physics and the experimental results on microscopic (atomic) systems, was brilliantly clarified by Heisenberg, who identified the basic crucial point which marks the deep philosophical difference between classical physics and the physics of microscopic (atomic) systems, briefly called *quantum systems*.

In the mathematical description of classical systems outlined in Sect. 1.2, it was recognized that dispersion free states are an unrealistic idealization (which would require infinite precision of measurements) and more correctly a state identified by realistic measurements defines a probability distribution on the random variables which describe the observables. Nevertheless, from the experience with classical macroscopic systems it was taken for granted that the ideal limit of dispersion free states could be approximated as closely as one likes, by refining the apparatus involved in the preparation and the identification of the state. This means that, e.g. even if one cannot prepare a state in which the q's and p's take a sharp value, one can prepare states ω in which the mean square deviations or variances $(\Delta_\omega q)^2$, $(\Delta_\omega p)^2$ are as small as one likes. This agrees with the assumption that the C^*-algebra of observables of a classical system is abelian (see also the discussion below). This philosophical prejudice seems strongly supported by experiments on macroscopic systems, but it is not so obvious when one deals with microscopic or atomic systems.

As a matter of fact, sharper and sharper measurements of the position of a particle, require instruments capable of distinguishing points at smaller and smaller scales. Now, for macroscopic systems, e.g. a material particle, to all effects it is enough to identify its position with a precision of a few orders of magnitude smaller than its size, so that for the realizability of the corresponding instruments one needs a control of the physics at scales which are still "macroscopic". The situation changes if one wants to localize the position of an atomic particle of size 10^{-8} cm or of a nucleus of size 10^{-13} cm. In fact, an accurate analysis of the operational ways of preparing a state of an atomic particle with sharper and sharper position and momentum shows that there are intrinsic limitations.

As argued by Heisenberg,[1] any attempt to measure the position of an atomic particle with sharper and sharper precision will produce a larger and larger disturbance on the microscopic system, with the result that the mean square deviation of the measurements of the momentum will become larger and larger. For example, a precise measurement of the position can be obtained by taking a "photograph", which requires sending light on the particle and the shorter the light wavelength λ is the sharper is the "picture" $((\Delta q)\,(\Delta \lambda^{-1}) \geq 1/4\pi)$. The picture is actually the result of a reflection of light by the particle and, since light rays carry energy and momentum, the reflection of light will change the particle momentum. It is an experimental fact that the shorter the wavelength is the larger is the recoil momentum of the particle, so that the smaller $(\Delta_\omega\, q)^2$ is the larger is $(\Delta_\omega\, p)^2$. The Einstein relation $p_{rad} = h/\lambda$, eq. (1.1.2), between the wavelength and the momentum of the radiation reflected by the particle, leads to the explicit bound $(\Delta_\omega\, q)\,(\Delta_\omega\, p) \geq h/4\pi$.

The above very sketchy, but important physical argument can be refined and extended to the various possible ways of measuring q and p. The result of such an analysis led Heisenberg to the conclusion that for any state ω there is an intrinsic limitation in the relative precision by which q and p can be measured, *independently of the state* ω. More precisely, if q_j, p_j denote the components of q and p in the j-th direction, the following limitation holds:

$$(\Delta_\omega\, q_j)\,(\Delta_\omega\, p_j)\ \geq h/4\pi \equiv \hbar/2. \qquad (2.1.1)$$

The above relations, called the *Heisenberg uncertainty relations*, should be regarded as inevitable limitations when one tries to measure the position and the momentum of a physical system and, therefore, when one tries to prepare a state with sharper and sharper values of position and velocity. Clearly, since h is very small, the above inequality is relevant only for microscopic systems and this is the crucial point where atomic physics departs from classical physics.

We now investigate the implications of Heisenberg uncertainty relations for the mathematical description of an atomic system. Given a state ω

---

[1]W. Heisenberg, *The Physical Principles of the Quantum Theory*, Dover 1930.

and any two observables $A = A^*$, $B = B^*$, which for simplicity can be taken of zero mean, we work out an algebraic lower bound on the product $\Delta_\omega (A) \Delta_\omega (B) = \omega (A^2)^{1/2} \omega (B^2)^{1/2}$.

Since $(A - i\lambda B)(A + i\lambda B) \geq 0$, $\forall \lambda \in \mathbf{R}$, positivity of ω implies

$$\omega (A^2) + |\lambda|^2 \omega (B^2) + i\lambda \omega ([A, B]) \geq 0, \qquad (2.1.2)$$

where $[A, B] \equiv AB - BA$. Since the last term is real, the positive-definiteness of the quadratic form in λ requires

$$4 \omega (A^2) \omega (B^2) \geq |\omega (i[A, B])|^2,$$

i.e.

$$\Delta_\omega (A) \ \Delta_\omega (B) \ \geq \ \frac{1}{2} |\omega ([A, B])|. \qquad (2.1.3)$$

Now, by Heisenberg analysis we know that uncertainty relations affect the mean square deviations of measurements of q_j and p_j and that this limitation is independent of the state; this strongly indicates, as realized by Heisenberg, that its roots must be looked for at the algebraic level. Indeed, since the general relations and properties of measurements must be encoded in the algebraic structure of the algebra of observables, which is *defined* in terms of measurements, the limitations in the relative precision of measurements should be read off in terms of algebraic relations. If the r.h.s. of eq. (2.1.3) is independent of ω, $[A, B]$ must be a multiple of the identity and Heisenberg idea is that the uncertainty relations arise as direct consequences of the following *Heisenberg commutation relations*

$$q_j \, p_k - p_k \, q_j = i\hbar \delta_{jk} \mathbf{1} \,, \qquad (2.1.4)$$

where δ_{jk} is the Kronecker symbol and the i is needed because $[q, p]^* = -[q, p]$. The choice with i replaced by $-i$ would be equally possible and the corresponding theory would be equivalent. Thus, the position and momentum of an atomic particle cannot be described by a commutative algebra.

Another example of uncertainty relations are those affecting the measurements of the orbital angle and of the angular momentum; more generally uncertainty relations occur for any pair of canonically conjugated variables.[2]

The deep philosophical conclusion out of the above discussion is that for the mathematical description of atomic systems one needs an *algebra of observables* which is *non-abelian*. Clearly, as always in the great physical discoveries, this is not a mathematical theorem and a great intuition and

---

[2]For an excellent review with reprints of the fundamental papers see J.A. Wheeler and W.H. Zureck, *Quantum Theory and Measurement*, Princeton University Press 1983; see also the references given in Sect. 1.1.

ingenuity was involved in Heisenberg foundations of Quantum Mechanics. To give up the abelian character of the algebra of observables may look as a very bold step, but it should be stressed that the commutativity of observables is a property of our mathematical description of classical *macroscopic* systems and it is not justified to extrapolate to the microscopic level the prejudices derived from our experience with the macroscopic world. The only guide must be the recourse to operational considerations and, as shown by Heisenberg, they indicate a non-abelian structure.

In conclusion, from the above considerations it follows that the right language for the mathematical description of quantum systems is the theory of (non-abelian) C^*-algebras and as such the mathematical structure of quantum mechanics can be viewed as a chapter of that theory.

2.2 States and representations. GNS construction

Having recognized that the observables of a quantum system generate a non-abelian C^*-algebra and that the states of the system are positive linear functionals on it, we now face the question of how such an abstract structure can be used for concrete physical problems, for calculations , predictions etc. We have therefore to find concrete realizations of the above structure. In the abelian case, thanks to the Gelfand-Naimark theory, we know that an abelian C^*-algebra is isometrically isomorphic to the algebra of continuous functions on its Gelfand spectrum and therefore in general abelian C^*-algebras are represented by algebras of continuous functions on a compact space. In the non-abelian case, it is not a priori obvious what are the concrete realizations of C^*-algebras. For this purpose we start with the following Definitions.

Definition 2.2.1 *A* **∗ homomorphism** *between two ∗-algebras* $\mathcal{A}$ *and* $\mathcal{B}$ *(with identities) is a mapping* $\pi : \mathcal{A} \rightarrow \mathcal{B}$*, which preserves all the algebraic relations including the ∗ , namely it is linear*

$$\pi(\lambda A + \mu B) = \lambda \pi(A) + \mu \pi(B), \quad \forall A, B \in \mathcal{A} \ , \ \lambda, \mu \in \mathbf{C},$$

∗ preserving, i.e. $\pi(A^*) = (\pi(A))^*$*, and multiplicative*

$$\pi(AB) = \pi(A)\,\pi(B), \quad \pi(\mathbf{1}_{\mathcal{A}}) = \mathbf{1}_{\mathcal{B}}.$$

If π *is one to one and onto (bijective), it is called a* **∗ isomorphism**. *A ∗ isomorphism of* $\mathcal{A}$ *into itself is called a* **∗ automorphism**.

Definition 2.2.2 *A* **representation** π *of a* C^*-*algebra* $\mathcal{A}$*, in a Hilbert space* $\mathcal{H}$*, is a ∗ homomorphism of* $\mathcal{A}$ *into the* C^*-*algebra* $\mathcal{B}(\mathcal{H})$ *of bounded (linear) operators in* $\mathcal{H}$.

A representation is **faithful** *if ker* $\pi = \{0\}$.

A representation is said to be **irreducible** *if* $\{0\}$ *and* $\mathcal{H}$ *are the only closed subspaces invariant under* $\pi(\mathcal{A})$.

Clearly, in an irreducible representation *every vector* $\Psi \in \mathcal{H}$ is *cyclic*, i.e. $\pi(\mathcal{A})\Psi$ is dense in $\mathcal{H}$. A representation π in a Hilbert space $\mathcal{H}$, with a cyclic vector Ψ shall be briefly denoted by $(\mathcal{H}, \pi, \Psi)$.

Proposition 2.2.3 *If* π *is a* * *homomorphism between two* C^*-*algebras* $\mathcal{A}$ *and* $\mathcal{B}$ *(with identities), then* $\forall A \in \mathcal{A}$

$$||\pi(A)||_{\mathcal{B}} \leq ||A||_{\mathcal{A}}. \qquad (2.2.1)$$

If π *is a* * *isomorphism the equality holds.*

Proof. We first consider the case $A = A^*$. Then, if $\lambda \mathbf{1} - A$ has an inverse,[3] so does $\pi(\lambda \mathbf{1} - A) = \lambda \mathbf{1} - \pi(A)$, so that the following inclusion of spectra holds:

$$\sigma_{\mathcal{B}}(\pi(A)) \subseteq \sigma_{\mathcal{A}}(A).$$

By the spectral radius formula applied to the abelian algebras generated by $\mathbf{1}$ and A, and by $\mathbf{1}$ and $\pi(A)$, respectively, one has

$$||\pi(A)||_{\mathcal{B}} = r_{\mathcal{B}}(\pi(A)) \leq r_{\mathcal{A}}(A) = ||A||_{\mathcal{A}},$$

where $r_{\mathcal{B}}$, $r_{\mathcal{A}}$ denote the spectral radius in $\mathcal{B}$ and in $\mathcal{A}$, respectively.

Then, for a generic $A \in \mathcal{A}$, we have

$$||\pi(A)||_{\mathcal{B}}^2 = ||\pi(A^*A)||_{\mathcal{B}} \leq ||A^*A||_{\mathcal{A}} = ||A||_{\mathcal{A}}^2.$$

If π is a * isomorphism, then it is invertible and by using eq. (2.2.1) for the inverse we get

$$||\pi(A)|| = ||A||.$$

As a first important structural result we prove that any state ω on a C^*-algebra $\mathcal{A}$ defines a representation π_ω of $\mathcal{A}$ in a Hilbert space $\mathcal{H}_\omega$, through the so-called *Gelfand-Naimark-Segal (GNS) construction*. The so obtained representation is briefly called the *GNS representation* defined by the state ω.

Theorem 2.2.4 *(GNS) Given a* C^*-*algebra* $\mathcal{A}$ *(with identity) and a state* ω, *there is a Hilbert space* $\mathcal{H}_\omega$ *and a representation* $\pi_\omega : \mathcal{A} \to \mathcal{B}(\mathcal{H}_\omega)$, *such that*
i) $\mathcal{H}_\omega$ *contains a cyclic vector* Ψ_ω,
ii) $\omega(A) = (\Psi_\omega, \pi_\omega(A)\Psi_\omega)$,

---

[3]Here and in the following discussion, for simplicity the identities in $\mathcal{A}$ and in $\mathcal{B}$ will be denoted by the same symbol $\mathbf{1}$.

iii) every other representation π in a Hilbert space $\mathcal{H}_\pi$ with a cyclic vector Ψ such that

$$\omega(A) = (\Psi, \pi(A)\Psi), \qquad (2.2.2)$$

is unitarily equivalent to π_ω, i.e. there exists an isometry $U : \mathcal{H}_\pi \to \mathcal{H}_\omega$ such that

$$U\pi(A)U^{-1} = \pi_\omega(A), \quad U\Psi = \Psi_\omega. \qquad (2.2.3)$$

Proof. $\mathcal{A}$ is a vector space and the state ω defines a semidefinite inner product on $\mathcal{A}$

$$(A, B) \equiv \omega(A^*B), \quad (A, A) = \omega(A^*A) \geq 0. \qquad (2.2.4)$$

The set

$$J \equiv \{A \in \mathcal{A} \ , \ \omega(B^*A) = 0, \quad \forall B \in \mathcal{A}\}$$

is a left ideal of $\mathcal{A}$, i.e. $\mathcal{A} \ J \subseteq J$. We can therefore consider the quotient space $\mathcal{A} \ /J$, the elements of which are the equivalence classes $[A] = \{A + B, \ B \in J\}$. The inner product induced on $\mathcal{A} \ /J$ by ω is well defined and strictly positive. The completion of $\mathcal{A} \ /J$ with respect to the topology induced by such an inner product is a Hilbert space $\mathcal{H}_\omega$.

To each $A \in \mathcal{A}$ we can associate an operator $\pi_\omega(A)$, acting on $\mathcal{H}_\omega$, by putting

$$\pi_\omega(A)[B] \equiv [AB], \qquad (2.2.5)$$

($\pi_\omega(A)$ is well defined since $[B] = [C]$ implies $\pi_\omega(A)[B] = \pi_\omega(A)[C]$, because J is a left ideal). Furthermore

$$||\pi_\omega(A)[B]||^2 = ([AB], [AB]) = \omega(B^*A^*AB) \equiv \omega_B(A^*A)$$

$$\leq \omega_B(\mathbf{1})||A||^2 = ||[B]||^2 \, ||A||^2,$$

since $\omega_B(A) \equiv \omega(B^*AB)$ is a positive linear functional and therefore continuous.[4] Thus

$$||\pi_\omega(A)|| \leq ||A|| \qquad (2.2.6)$$

and, since $\pi_\omega(A)^*[B] = \pi_\omega(A^*)[B]$, π_ω is a * homomorphism of $\mathcal{A}$ into $\mathcal{B}(\mathcal{H}_\omega)$.

With the identification $\Psi_\omega \equiv [\mathbf{1}]$, we get $\pi_\omega(A)\Psi_\omega = [A]$, so that Ψ_ω is cyclic. Finally,

$$(\Psi_\omega, \pi_\omega(A)\Psi_\omega) = ([\mathbf{1}], [A]) = \omega(\mathbf{1}^*A) = \omega(A) \ . \qquad (2.2.7)$$

The isometry U is defined by

$$U^{-1}\pi_\omega(A)\Psi_\omega = \pi(A)\Psi.$$

In fact, U is norm preserving by eq. (2.2.2) and its range is dense, so that it has a unique extension to a unitary map from $\mathcal{H}_\omega$ to $\mathcal{H}$. The last equation

---

[4]See Proposition 1.6.3 of Appendix C.

of the theorem is easily proved on the dense set $\pi(\mathcal{A})\Psi$ and therefore in $\mathcal{H}$.

The GNS construction is very important from a general mathematical point of view, since it reduces the existence of Hilbert space representations of a C^*-algebra to the existence of states, which is guaranteed by Proposition 1.6.4 of Appendix C.

It is also important for the implications on the description of physical systems discussed in Sect. 1.3, since it says that the (experimental) set of expectations of the observables given by a state have a Hilbert space interpretation in terms of a representation of the observables by Hilbert space operators and the description of the state (expectations) in terms of matrix elements of a Hilbert space vector. Thus, the basis of the mathematical description of quantum mechanical systems need not to be postulated, as in the Dirac-Von Neumann axiomatic setting of quantum mechanics, [5] but it is merely a consequence of the C^*-algebra structure of the observables argued in Sect.1.3 and of the fact that, by its operational definition, a state defines a positive linear functional on them. If the algebra $\mathcal{A}$ is abelian the GNS representation defined by a (faithful) state ω is equivalently described by the probability space $(\Sigma(\mathcal{A}), \Sigma_{\mu_\omega}, \mu)$, with Σ_{μ_ω} the family of μ_ω-measurable subsets of $\Sigma(\mathcal{A})$ and μ_ω the Riesz-Markov measure defined by ω (see Sect. 2.4).

Given a state ω, the vectors of the representation space $\mathcal{H}_\omega$ define states on $\mathcal{A}$ since $\forall \Phi \in \mathcal{H}_\omega$, $\varphi(A) \equiv (\Phi, \pi_\omega(A)\Phi)$ is a positive linear functional on $\mathcal{A}$. The so obtained states have the physical interpretation of states obtained by "acting" on ω by observables, i.e. by physically realizables operations, since, by the GNS construction, vectors of the form $\pi(A)\Psi_\omega$, $A \in \mathcal{A}$ are dense in $\mathcal{H}_\omega$.

Thus, the GNS construction gives a mapping between states and Hilbert space vectors, correspondingly called *state vectors*, so that the observables are represented by operators and the expectations $\varphi(A)$ are given by the "matrix elements" $(\Phi, \pi_\omega(A)\Phi)$.

Clearly, if $\Phi \in \mathcal{H}_\omega$ is cyclic, the representation $(\mathcal{H}_\omega, \pi_\omega, \Phi)$ is unitarily equivalent to the GNS representation defined by the state $\varphi(A) \equiv (\Phi, \pi_\omega(A)\Phi)$, since the two representations are related as in iii) of the GNS Theorem above.

Quite generally, two representations of a C^*-algebra are *unitarily equivalent* if they are related by an isometry as in eq. (2.2.3). In general, given any two states ω_1, ω_2, the corresponding GNS representations need not to be unitarily equivalent.

---

[5]See e.g. P.A.M. Dirac, *The Principles of Quantum Mechanics*, Oxford Claredon Press 1958.

Given any two states ω_1, ω_2 the convex linear combination

$$\omega \equiv \lambda\omega_1 + (1-\lambda)\omega_2, \quad 0 < \lambda < 1, \tag{2.2.8}$$

is also a (normalized) state, called a *mixture* of the states ω_1, ω_2, or briefly a *mixed state*, and ω majorizes the functionals $\lambda\omega_1$ (i.e. $\omega - \lambda\omega_1$ is positive) and $(1-\lambda)\omega_2$. A state ω is called a *pure state* if it cannot be written as a convex linear combination of other states, equivalently if the only positive linear functionals majorized by ω are of the form $\lambda\omega$, $0 < \lambda < 1$.

One can show that the GNS representation defined by a state ω is irreducible iff ω is pure (Appendix D). Thus, a mixed state ω cannot be represented by a state vector of an irreducible representation; rather, e.g. it is represented by a *density matrix* in an irreducible representation π in a Hilbert space $\mathcal{H}$, i.e. there exists a *positive trace-class operator* ρ in $\mathcal{H}$, (i.e. such that $\mathrm{Tr}\,|\rho| < \infty$), with $\mathrm{Tr}\rho = 1$, such that

$$\omega(A) = Tr\,(\rho\,\pi(A)), \quad \forall A \in \mathcal{A}\ . \tag{2.2.9}$$

Given a representation π, the set of states of the above form (which include as a special case the state vectors, when ρ is a one-dimensional projection), is called the *folium of the representation π*.

A state ω is *faithful* if $\omega\,(A^*A) > 0, \forall A \neq 0$ and clearly the corresponding GNS representation is faithful.

2.3   Gelfand-Naimark theorem: observables as operators

By the GNS theorem, every state defines a concrete realization of a C^*-algebra as operators in a Hilbert space, but this realization may not be faithful and one would like to characterize the faithful representations in general. This is the content of the Gelfand-Naimark (GN) theorem, by which every C^*-algebra is isomorphic to an algebra of bounded operators in a Hilbert space, the vectors of which describe a full set of states.

Thus, the GN theorem can be read as the derivation of two basic Dirac-Von Neumann axioms (about the representations of states and observables in quantum mechanics) merely from the C^*-algebraic structure of the observables. Such a general Hilbert space description is equivalent to a representation in terms of continuous functions and probability measures only if the algebra of observables is abelian (see below). The GN theorem is therefore very important for the mathematical description of a physical system, because it settles the basic difference between classical and quantum physics.

Theorem 2.3.1 *(Gelfand-Naimark) A C^*-algebra $\mathcal{A}$ is isomorphic to an algebra of (bounded) operators in a Hilbert space.*

Proof. Let $\mathcal{F}$ be a family of states on $\mathcal{A}$, with the property that they separate $\mathcal{A}$, [6] and consider the direct sum [7] of the GNS representations defined by $\omega \in \mathcal{F}$

$$\mathcal{H} = \oplus_{\omega \in \mathcal{F}} \mathcal{H}_\omega , \quad \pi = \oplus_{\omega \in \mathcal{F}} \pi_\omega , \quad \text{i.e.} \quad [\pi(A) x]_\omega = \pi_\omega(A) x_\omega.$$

Clearly, since $\forall \omega \in \mathcal{F}$, $||\pi_\omega(A)|| \le ||A||$, the direct sum $\pi(A)$ is a bounded operator in $\mathcal{H}$

$$||\pi(A) x||^2 = \sum_\omega ||\pi_\omega(A) x_\omega||^2 \le ||A||^2 \sum_\omega ||x_\omega||^2 = ||A||^2 ||x||^2$$

and π is a representation of $\mathcal{A}$ into $\mathcal{B}(\mathcal{H})$. Since the family $\mathcal{F}$ is separating, $\forall A \in \mathcal{A}$, $A \ne 0$, there exists at least one ω such that $\pi_\omega(A) \ne 0$ and therefore $\pi(A) \ne 0$. Hence, $\ker \pi = \{0\}$, i.e. π is a * isomorphism between $\mathcal{A}$ and the C^*-algebra $\pi(\mathcal{A})$ and, by Proposition 2.2.3,

$$||\pi(A)|| = ||A||.$$

It is an instructive exercise to get the Gelfand-Naimark characterization of abelian C^*-algebras from the above theorem. In fact, by Proposition 2.6.2 the irreducible representations π_ω are defined by pure states, which for abelian C^*-algebras are multiplicative (Proposition 2.6.3), so that the corresponding representation are one dimensional $\pi_\omega(A) = \omega(A)\mathbf{1}$. Then the family $\mathcal{F}$ of all inequivalent irreducible representations coincides with the Gelfand spectrum and the faithful representation $\pi(A) = \oplus_{\omega \in \mathcal{F}} \pi_\omega(A)$ is given by the collection $\{\omega(A), \omega \in \mathcal{F}\}$, i.e. by the function $\tilde{A}(\omega) \equiv \omega(A)$. Furthermore,

$$||\tilde{A}||_\infty = \sup_\omega |\tilde{A}(\omega)| = \sup_\omega |\omega(A)| = ||A||.$$

With the weak* topology $\mathcal{F}$ is a compact Hausdorff topological space by the Alaoglu-Banach theorem and the functions $\tilde{A}$ are continuous (by the same

---

[6] The existence of such a family is guaranteed by construction in our approach, as discussed is Sect. 1.3, and it is actually true in general for C^*-algebras as proved in Proposition 1.6.5.

[7] We recall that given a family $\mathcal{H}_\alpha$, $\alpha \in I$, of Hilbert spaces, one defines the direct sum $\mathcal{H} = \oplus_{\alpha \in I} \mathcal{H}_\alpha$ as the set of vectors of the form $x = \{x_\alpha, \alpha \in I\}$, with $\sum ||x_\alpha||^2 < \infty$. The scalar product

$$(\{x_\alpha\}, \{y_\alpha\}) \equiv \sum (x_\alpha, y_\alpha)$$

is well defined since

$$|\sum (x_\alpha, y_\alpha)| \le \sum |(x_\alpha, y_\alpha)| \le \sum ||x_\alpha|| \, ||y_\alpha||$$

$$\le (\sum ||x_\alpha||^2)^{1/2} (\sum ||y_\alpha||^2)^{1/2} < \infty.$$

Similarly one proves that $\{x_\alpha\} + \{y_\alpha\} \in \mathcal{H}$ and that $\mathcal{H}$ is complete.

arguments as in the last part of the proof of Theorem 1.5.5). This approach shows one basic difference between the abelian and the non abelian case. In the first case, the set of pure states defines a "classical" space; in the second case, the set of pure states defines a "quantum" or "non-commutative" space, whose points are the Hilbert spaces of the GNS representations defined by the pure states. Whereas in the first case each representation is one dimensional and each element $A \in \mathcal{A}$ can only take one value, in the second case in each representation, i.e. in each point of the non commutative space, an element A acts as an operator.

2.4   The probabilistic interpretation. * Quantum probability

Given a normal element $A \in \mathcal{A}$, one may consider the abelian algebra $\mathcal{A}_A$ generated by A, A^* and $\mathbf{1}$, the Gelfand spectrum of which coincides with the spectrum $\sigma(A)$ of A; then, by the Riesz-Markov theorem, a state ω on $\mathcal{A}$ defines a probability measure $\mu_{\omega,A}$ on $\sigma(A)$, so that $\forall B \in \mathcal{A}_A$

$$\omega\left(B\right) = \int_{\sigma(A)} \tilde{B}(\lambda)\, d\mu_{\omega,A}(\lambda), \qquad (2.4.1)$$

with $\tilde{B}(\lambda)$ the Gelfand transform of B.

In this way, for any given observable the probabilistic interpretation of a state follows quite generally from its being a positive functional, exactly as in the classical theory (see Sect. 1.2). [8] In the non abelian case, however, one meets substantial differences: the probability measure is not the same for all the observables; a state defines one probability measure for each abelian algebra generated by a normal element of $\mathcal{A}$, but not a joint probability distribution for all observables. This structure has been taken as a prototype of what is nowadays called Quantum Probability.[9]

---

[8]We do not enter into the problem of the description of the interaction between the (quantum) system and the instrument (the so-called *measurement problem*) and the related "reduction of the wave packet", because a detailed theory of such an interaction is still under discussion and in fact goes beyond the structure discussed in these lectures. Arguments for the decoherence of classical apparatuses as a consequence of the their macroscopic character have been discussed in K. Hepp, Helv. Phys. Acta **45**, 237 (1972); the "reduction of the wave packet" amounts to the instability of macroscopic systems against their existence as mixtures, leading to sharp macroscopic states. For a beautiful review of these problems see A.S. Wightman, Some comments on the quantum theory of measurement, in *Probabilistic Methods in Mathematical Physics*, F. Guerra et al. eds., World Scientific 1992.

[9]See e.g. S.P. Gudder, *Quantum Probability*, Academic Press 1988; P.A. Meyer, *Quantum Probability for Probabilists*, Springer 1993; P. Biane and R. Durrett, *Lectures on Probability Theory*, Springer 1994.

To clarify the relation and differences between classical and quantum probability we briefly recall the basic ideas of classical probability theory.

The underlying structure is the abstract measure theory, which generalizes Lebesgue theory. The latter one is based on the triplet $(\mathbf{R}, \mathcal{B}, \mu)$, where $\mathcal{B}$ is the family of Borel sets of $\mathbf{R}$, namely the smallest family of subsets closed under complementation and countable unions and containing each open interval, and μ is the Lebesgue measure defined on each open interval (a, b) by $\mu((a, b)) = |b - a|$ and having a unique σ-additive Borel regular [10] extension to $\mathcal{B}$.

In strict analogy, the abstract measure theory is based on the following structure

i) a set Ω and a family Σ of subsets of Ω, closed under complementation and countable unions with $\Omega \in \Sigma$, (i.e. a σ-field of subsets of Ω), called the *measurable sets*. The pair (Ω, Σ) is called a *measurable space*.

ii) a *measure* μ on Σ, i.e. a map $\Sigma \to \mathbf{R}_+ \cup \infty$ with the property that
a) $\mu(\emptyset) = 0$,
b) $\mu(\cup_i A_i) = \sum_i \mu(A_i)$, if $A_i \cap A_j = \emptyset$, $\forall i \neq j$.

For technical reasons and for simplicity, we shall also assume that μ is σ finite, i.e.
c) $\Omega = \cup_i A_i$, with $\mu(A_i) < \infty$.

A *measurable function* $f : \Omega \to \mathbf{R}$ is a function with the property that $\forall B \in \mathcal{B}$ (the family of Borel sets of $\mathbf{R}$), $f^{-1}(B) \in \Sigma$. Clearly, in this case one can define the integral of f

$$\int_\Omega f \, d\mu$$

as in the ordinary Lebesgue case.

The theory of probability is the mathematical description of *random variables*, i.e. of functions of events which have a certain probability of occurrence. To make these concepts more precise, one considers the *set Ω of events*, which may occur.

A few basic operations are naturally defined on events
a) if A and B are two events, the *union* $A \cup B$ is the event which occurs if at least one of the two events occurs
b) the *complement* A^c of the event A is the event which occurs if A does not occur

---

[10] A σ-additive measure is Borel regular if it is defined on all Borel sets and has the following property: $\forall S \in \mathcal{B}$

$$\mu(S) = \sup_C \{\mu(C); C \subset S, \ C \text{ compact and Borel}\}$$
$$= \inf_O \{\mu(O); S \subset O, O \ \text{open}\}.$$

c) the *intersection* $A \cap B$ is the event which occurs if both A and B occur

d) the impossible event is denoted by $\emptyset$ and clearly $\Omega \equiv$ the union of all events, is the certain event.

One easily recognizes in the above structure the basic features of the propositional calculus (see below), i.e. the events form a Boolean algebra of sets. In conclusion, the basic elements of the theory of probability are a set Ω and a family Σ of subsets of Ω, closed under complementations and unions, with $\emptyset, \Omega \in \Sigma$.

A *probability* is defined on Σ by assigning a positive function $\mu : \Sigma \to [0, 1]$ with the obvious meaning that $\mu(A)$, $A \in \Sigma$ is the probability that the event A occurs. It is rather clear that μ must have the following properties in order to describe the probability of occurrence of events

a) $\mu(\emptyset) = 0$, $\mu(\Omega) = 1$,

b) if $A \cap B = \emptyset$, $\mu(A \cup B) = \mu(A) + \mu(B)$.

From an operational point of view, the probability $\mu(A)$ can be defined as the limit of the frequency $\nu_N(A)$ of occurrence of the event A, during N observations or trials, performed on the given system, or on replicas of it, when the number N goes to infinity

$$\mu(A) = \lim_{N \to \infty} \nu_N(A).$$

As realized by Kolmogorov, [11] the theory of probability becomes naturally embedded in the abstract measure theory if the family of events Σ is closed under countable unions and the probability μ is required to be σ-additive; hence μ becomes a *probability measure*. It is not difficult to justify these requirements on the basis of operational considerations.[12] Then, the triple (Ω, Σ, μ) is called a *probability space*.

A *random variable* f is a measurable function on Ω and its *expectation* with respect to the probability measure μ is defined by

$$< f >_\mu \equiv \int_\Omega f \, d\mu.$$

The *probability distribution function* $\mu_f(x)$, $x \in \mathbf{R}$ associated to the random variable f is the measure on $\mathbf{R}$ defined by

$$\mu_f(B) \equiv \mu(f^{-1}(B)), \quad \forall B \in \mathcal{B}.$$

Clearly, each probability measure μ assigns a probability to each event and it defines a positive linear functional on the algebra $\mathcal{A}$ of bounded random variables

$$\omega_\mu(f) = \int_\Omega f d\mu = \int_\mathbf{R} x \, d\mu_f(x).$$

---

[11] A.N. Kolmogorov, *Foundations of the Theory of Probability*, Chelsea 1956.

[12] See e.g. P.R. Halmos, *Measure theory*, Van Nostrand 1950, Chap. IX.

From this point of view, the basic difference between classical and quantum probability is that in the latter case the algebra of observables $\mathcal{A}$ cannot be realized as an algebra of random variables on a probability space; such a realization is possible for each commutative subalgebra of $\mathcal{A}$, but the probability spaces corresponding to two non-commuting observables are different.

This basic difference can be traced back to the different structure of classical and quantum events, where the events can be given the physical interpretation of outcomes of trials or of experiments. By relying on the logical structure of propositions (for more details see the next Section), in terms of which the events can be described, one can argue in general [13] that the family of events has the structure of an orthocomplemented lattice with a smallest or zero element $\emptyset$ and a largest or unit element $\mathbf{1}$.

For the convenience of the reader, we briefly recall that a *lattice* is a set L with
1) a *partial order* relation $\leq$, (i.e. a binary relation satisfying $\forall a, b, c, \in L$: (reflexivity) $a \leq a$; (antisymmetry) $a \leq b$, $b \leq a \Rightarrow a = b$; (transitivity) $a \leq b$, $b \leq c \Rightarrow a \leq c$),
2) two binary operations $\vee$ and $\wedge$, called *join* and *meet*, satisfying

$$a \leq a \vee b, \quad b \leq a \vee b, \quad (a \leq c, b \leq c) \Rightarrow a \vee b \leq c,$$

$$a \wedge b \leq a, \quad a \wedge b \leq b, \quad (c \leq a, c \leq b) \Rightarrow c \leq a \wedge b.$$

The join and meet operations are *commutative* ($a \vee b = b \vee a$, etc.), *associative* ($a \vee (b \vee c) = (a \vee b) \vee c$ etc.), *idempotent* ($a \vee a = a, b \wedge b = b$) and *absorptive* (i.e. $a \vee (a \wedge b) = a \wedge (a \vee b) = a$). Moreover $a \leq b$, $a \wedge b = a$, $a \vee b = b$ are all equivalent relations.

The partial order relation may be given the meaning of inclusion or of "smaller than or equal to", the join operation corresponds to the supremum or the least upper bound or to the set theoretical union, the meet corresponds to the infimum or to the greatest lower bound or to the set intersection.[14] Thus, e.g. given two events a, b, the order relation $a \leq b$ has the meaning that a is included in b, $a \vee b$ is the event corresponding to (a *or* b), $a \wedge b$ is the event corresponding to (a *and* b) etc.

A lattice is *complete* if it contains a smallest or *zero* element $\emptyset$ and a largest or *unit* element $\mathbf{1}$, characterized by $\emptyset \leq a$, $a \leq \mathbf{1}$, $\forall a \in L$. A lattice is *orthocomplemented* if $\forall a \in L$ there exists an element a', called

---

[13]G. Birkhoff and J. von Neumann, Ann. Math. **37**, 823 (1936); G. Mackey, *Mathematical Foundations of Quantum Mechanics*, Benjamin 1963, esp. Sect. 2.2; J.M. Jauch, *Foundations of Quantum Mechanics*, Addison Wesley 1968, esp. Chap. 5; C. Piron, *Foundations of Quantum Physics*, Benjamin 1976, esp. Chap. 2.

[14]For more details see e.g. G. Birkhoff, *Lattice Theory*, Am. Math. Soc. 1963; P.R. Halmos, *Lectures on Boolean Algebras*, Van Nostrand 1963.

the complement of a, such that $a \vee a' = 1$, $a \wedge a' = \emptyset$, $(a')' = a$, and $a \leq b \Rightarrow b' \leq a'$. A lattice is *distributive* if

$$a \wedge (b \vee c) = (a \wedge b) \vee (a \wedge c), \tag{2.4.2}$$

equivalently if $a \vee (b \wedge c) = (a \vee b) \wedge (a \vee c)$. A complete orthocomplemented distributive lattice with zero and unit elements is called a *Boolean algebra*.[15]

It is easy to check that the σ-field Σ of subsets of a set Ω is a Boolean algebra and it is a deep result by Stone that any Boolean algebra is isomorphic to a lattice of subsets of a set, with the join and meet operations corresponding to unions and intersections, the complement to the set complementation, $\emptyset$ to the empty set and $\mathbf{1}$ to the union of all lattice elements.

It is also not difficult to justify that the outcomes of trials or of experiments (typically parametrized by the yes/no outputs) have a lattice structure,[16] but distributivity is a crucial property, which is satisfied by classical events, but not by quantum events. For example, in the double slit interference experiment (discussed in Sect. 5.1) the quantum events $a=$interference pattern produced by the particle on the screen, $b=$ the particle has passed through the slit 1, $b'=$ the particle has passed through the slit 2, do not satisfy lattice distributivity since a is incompatible with both b and b' (the interference pattern does not appear if the particle passes through one of the slits!), so that

$$\emptyset = a \wedge b = a \wedge b' = (a \wedge b) \vee (a \wedge b'),$$

whereas

$$a \wedge (b \vee b') = a \wedge \mathbf{1} = a.$$

As a consequence of this lack of distributivity, the lattice of quantum events is not isomorphic to a family of subsets of a set as in the classical case, where classical probability theory can be applied, but rather to a family of orthogonal projections on a Hilbert space $\mathcal{H}$, with the partial order relation given by the projection inclusion ($P \leq Q$ iff $PQ = QP = P$), the join $P \vee Q$ given by the orthogonal projection on the span of $P\mathcal{H} \cup Q\mathcal{H}$, the meet $P \wedge Q$ by the orthogonal projection on $P\mathcal{H} \cap Q\mathcal{H}$ and the complement P' by $\mathbf{1} - P = P^{\perp}$. It is easy to see that lattice distributivity fails if the projections do not commute.

Another way of discussing the structural difference between classical and quantum probability is by realizing that a classical probability space can

---

[15] In a Boolean algebra $\mathcal{B}$ one may define a difference $a - b \equiv a \wedge b'$ and an addition modulo 2 or symmetric difference $a + b = (a' \wedge b) \vee (a \wedge b')$. Then $\mathcal{B}$ with $+$ as addition and $\wedge$ as multiplication is an algebraic commutative ring with unit $\mathbf{1}$ and zero $\emptyset$, over the field $K = \{0, 1\}$; since $a + a = \emptyset$, every element is idempotent and $\mathcal{B}$ is of characteristic 2.

[16] See e.g. the above references to Birkhoff and von Neumann, to Jauch and to Piron.

equivalently be described by a triple $(\mathcal{H}, \mathcal{P}, \psi)$, where $\mathcal{H}$ is a Hilbert space, $\mathcal{P}$ a set of commutative orthogonal projections in $\mathcal{H}$ and ψ a vector of $\mathcal{H}$. In fact, given a triplet (Ω, Σ, μ), one can consider the quotient $\tilde{\Sigma} = \Sigma / \mathcal{J}_\mu$, $\mathcal{J}_\mu \equiv \{A \in \Sigma, \mu(A) = 0\}$ and the C^*-algebra $\mathcal{A}$ of bounded measurable functions on Ω generated by the characteristic functions of the sets of $\tilde{\Sigma}$. Then μ defines a faithful state ω_μ on $\mathcal{A}$ by

$$\omega_\mu(f) = \int_\Omega f(x) \, d\mu(x),$$

$(\omega_\mu(|f|^2) = 0$ implies that f is equivalent to the zero element), and by the GNS theorem ω_μ defines a representation of $\mathcal{A}$, in a Hilbert space $\mathcal{H}$, with cyclic vector Ψ_1. This representation is clearly isomorphic to the representation given by $L^2(\Omega, \mu)$, with the elements of $\mathcal{A}$ represented by multiplication operators and the cyclic vector given by an element $\psi(x) \in \mathcal{H}$ with spectral measure $d\mu(x)$. Clearly the algebra $\mathcal{A}$ is generated by the family $\mathcal{P}$ of orthogonal commutative spectral projections. In conclusion we have constructed the triplet $(\mathcal{H}, \mathcal{A}, \psi)$.

Conversely, given an abelian C^*-algebra of operators with a cyclic faithful state Ψ_1, by the Riesz-Markov theorem there exists a spectral measure μ on the spectrum Ω of $\mathcal{A}$, such that $\forall A \in \mathcal{A}$

$$(\Psi_1, A \, \Psi_1) = \int_\Omega \hat{A}(x) \, d\mu(x),$$

with $\hat{A}$ a bounded multiplication operator, i.e. we recover the triplet (Ω, Σ, μ), where Σ is the family of μ-measurable sets of Ω, modulo the sets of zero measure.

In this logic, quantum probability is defined by a triplet $(\mathcal{H}, \mathcal{A}, \psi)$, where $\mathcal{H}$ is a Hilbert space, $\mathcal{A}$ is the C^*-algebra of observables (in general non-commutative) and ψ is a state on $\mathcal{A}$; equivalently, instead of $\mathcal{A}$ one may use the set $\mathcal{P}$ of orthogonal projections on $\mathcal{H}$ which generate the observables. Whereas in the commutative case the family $\mathcal{P}$ has the structure of a Boolean algebra, in the quantum non-commutative case it has the structure of an orthocomplemented *non-distributive* lattice. Also a state is a more general concept than a measure; a state ω on $\mathcal{A}$ defines a measure $\mu_{\omega, \mathcal{A}_1}$ for each abelian subalgebra $\mathcal{A}_1$ of $\mathcal{A}$, but not a single measure for $\mathcal{A}$. As in the case of a probability measure, a state defines a positive linear functional on the observable algebra, but subadditivity fails in general:

$$\omega(P_1 \vee P_2) \not\leq \omega(P_1) + \omega(P_2),$$

if the projections P_1, P_2 do not commute.

2.5   * Quantum logic

The mathematical structure at the basis of classical logic is that of a Boolean algebra, i.e. of an orthocomplemented complete lattice satisfying the distributive law, and therefore by Stone theorem it is isomorphic to a lattice of subsets of a set, which corresponds to a commutative algebra. Quantum logic is the generalization of classical logic in which lattice distributivity does not hold and consequently it corresponds to a noncommutative algebra. For the convenience of the reader, a brief sketch of the above structures is outlined below; for more details we refer to the references given in the Introduction.

The aim of *mathematical logic* is a formal theory of reasoning and the basic structure is the propositional calculus, by which simple propositions (henceforth denoted by latin letters $a, b, ...$) can be combined and the truth or falsity of the resulting assertion is formalized in terms of basic mathematical operations:

1) the *equivalence* of two propositions a, b is denoted by $a = b$,

2) the *negation* or complement of a proposition a is denoted by a' (a is true iff a' is false),

3) the *disjunction* of two propositions a, b is denoted by $a \vee b$ (it corresponds to the assertion "a *or* b", in the non-exclusive sense),

4) the *conjunction* is denoted by $a \wedge b$ (it corresponds to the assertion "a *and* b"),

5) the *implication* of b by a is denoted by $a \leq b$ (equivalently one may write $b \geq a$) and can be characterized by $a = a \wedge b$, equivalently by $a \vee b = b$.

From the logical interpretation one has that the operations $\vee, \wedge$ are commutative, associative, idempotent and absorptive (see the previous Section) and that the implication is a partial order relation, (i.e. reflexive, antisymmetric and transitive). Furthermore one has that

$$a \leq a \vee b; \quad a \leq c, b \leq c, \Rightarrow a \vee b \leq c,$$

$$a \geq c, b \geq c, \Rightarrow a \wedge b \geq c,$$

since if a is true also "a or b" is true etc. Similarly one has

$$(a')' = a; \quad a \leq b \Rightarrow b' \leq a',$$

$$(a \wedge b)' = a' \vee b'; \quad (a \vee b)' = a' \wedge b'.$$

The disjunction of all propositions is implied by any proposition and it has the meaning of the trivial proposition, since it is always true; it will be denoted by $\mathbf{1}$. Similarly, the conjunction of all propositions implies all propositions, it has the meaning of the absurd proposition, since it implies a and the negation of a, it is always false and it is denoted by $\emptyset$. Clearly

$$a \vee a' = \mathbf{1}, \quad a \wedge a' = \emptyset.$$

Thus one has the structure of an orthocomplemented complete lattice as defined in the previous Section.

The crucial characteristic property of classical mathematical logic is the distributivity of the operations $\vee, \wedge$:

$$a \wedge (b \vee c) = (a \wedge b) \vee (a \wedge c),$$

$$a \vee (b \wedge c) = (a \vee b) \wedge (a \vee c).$$

The meaning is rather transparent in terms of its logical interpretation; e.g. the first equation says that "a *and* (b *or* c)" is equivalent to "(a *and* b) *or* (a *and* c)". The implications are worthwhile to be remarked; for example, choosing $c = b'$ in the first equation, one has that any proposition a can be "decomposed" along b and b':

$$a = (a \wedge b) \vee (a \wedge b'),$$

i.e. the proposition a is equivalent to "(a *and* b) *or* (a *and* *not-b*)".

Thus, classical mathematical logic has the structure of a Boolean algebra and by Stone theorem is isomorphic to a lattice of subsets of a set.

As stressed by Birkhoff and Von Neumann the lattice of propositions for quantum systems does not satisfy distributivity. Indeed if a and b are not compatible propositions, e.g. yes/no outputs of experiments corresponding to non-compatible (i.e. non-commuting) observables, then $a \wedge b = \emptyset = a \wedge b'$ and the decomposition of a derived above from distributivity cannot hold. For an explicit example, see e.g. the case of the double slit interference discussed in the previous Section. Classical mathematical logic applies only to propositions which refers to compatible, i.e. commuting, observables.

2.6   Appendix D: States and representations

For the convenience of the reader, we present a few useful results on states and representations. For a deeper insight the reader is referred to the textbooks on C^*-algebras listed in the Introduction and in Sect. 1.3.

Proposition 2.6.1 *A representation π of a C^*-algebra $\mathcal{A}$ in a Hilbert space $\mathcal{H}$ is irreducible iff the commutant*

$$\pi(\mathcal{A})' \equiv \{ C \in \mathcal{B}(\mathcal{H}), \ [C, \pi(A)] = 0, \ \forall A \in \mathcal{A} \} \qquad (2.6.1)$$

consists only of multiples of the identity.

Proof. If $\exists C \in \pi(\mathcal{A})'$, which is not a multiple of the identity, so is C^* and we may take C self-adjoint $(C \to C + C^*, i(C - C^*))$. Then, its spectral projections define non-trivial subspaces of $\mathcal{H}$ invariant under $\pi(\mathcal{A})$, i.e. $\pi(\mathcal{A})$ is reducible.
Conversely, if $\pi(\mathcal{A})$ is reducible, there exists a non trivial (orthogonal) projection $P \neq \mathbf{1}$ on an invariant subspace and $[P, \pi(\mathcal{A})] = 0$; then the commutant cannot consists only of multiples of the identity.

Proposition 2.6.2 *The GNS representation defined by a state ω is irreducible iff ω is pure.*

Proof. If the representation is reducible, let $P \neq \mathbf{1}$ be the projection on a non-trivial invariant subspace. Then $P\Psi_\omega \neq 0$, $(\mathbf{1} - P)\Psi_\omega \neq 0$, because Ψ_ω is cyclic, and $P\pi(\mathcal{A})(\mathbf{1} - P) = 0 = (\mathbf{1} - P)\pi(\mathcal{A})P$, because P commutes with $\pi(\mathcal{A})$. Thus

$$\omega(A) = (\Psi_\omega, \pi_\omega(A)\Psi_\omega)$$

$$= (P\Psi_\omega, \pi_\omega(A)P\Psi_\omega) + ((\mathbf{1} - P)\Psi_\omega, \pi_\omega(A)(\mathbf{1} - P)\Psi_\omega)$$

$$\equiv \lambda\omega_1(A) + (1 - \lambda)\omega_2(A), \quad \lambda = ||P\Psi_\omega||^2 = (\Psi_\omega, P\Psi_\omega).$$

Since the so defined ω_1, ω_2 are states, ω cannot be pure.
Conversely, if π_ω is irreducible, suppose that ω is a mixture of the states ω_1, ω_2. Then, $\exists \lambda \in (0,1)$ such that ω majorizes $\lambda\omega_1$ and therefore $\lambda\omega_1(A^*B)$ defines a bounded sesquilinear form on $\mathcal{H}_\omega$,

$$\lambda\omega_1(A^*A) \leq \omega(A^*A) \leq ||A||^2$$

and, by the Riesz representation theorem, there exists a bounded operator T on $\mathcal{H}_\omega$, such that, $\forall A, B, C \in \mathcal{A}$, $\lambda\omega_1(A^*B) = (\Psi_\omega, \pi(A^*)T\pi(B)\Psi_\omega)$ and

$$\omega(A^*TCB) = \lambda\omega_1(A^*CB) = \lambda\omega_1((C^*A)^*B)$$

$$= \omega((C^*A)^*TB) = \omega(A^*CTB).$$

This implies that $T \in \pi_\omega(\mathcal{A})'$, in contrast with the irreducibility, unless $T = \mu \mathbf{1}$, $\mu \in \mathbf{R}^+$ and $\mu \omega = \lambda \omega_1$.

It follows easily from the above Proposition that the state vectors of an irreducible representation define pure states.

Proposition 2.6.3 *A state ω on a commutative C^*-algebra $\mathcal{A}$ (with identity) is pure iff it is multiplicative, i.e.*

$$\omega\,(AB) = \omega(A)\ \ \omega\,(B). \tag{2.6.2}$$

Proof. Eq. (2.6.2) is equivalent to

$$\pi_\omega(\mathcal{A}) = \{\lambda \mathbf{1}\,,\ \lambda \in \mathbf{C}\},$$

which is implied by irreducibility since for abelian algebras $\pi_\omega(\mathcal{A}) \subseteq \pi_\omega(\mathcal{A})'$ and conversely implies that the GNS representation space is one dimensional so that $\pi_\omega(\mathcal{A})' = \{\lambda \mathbf{1}\,,\ \lambda \in \mathbf{C}\}$ and irreducibility follows.

If $A = A^*$ and $\lambda \in \sigma(A)$, then $\exists\ \omega$ such that $\omega(A) = \lambda$; if ω is dispersion free on A, i.e. $\Delta_\omega(A) = 0$, then, in the GNS representation π_ω, λ is an eigenvalue and Ψ_ω is a corresponding eigenstate, since

$$0 = \Delta_\omega(A)^2 = ||(A - \lambda \mathbf{1}\,)\,\Psi_\omega||^2 \ \ \Rightarrow\ \ A\,\Psi_\omega = \lambda\,\Psi_\omega.$$

2.7   * Appendix E: Von Neumann algebras

As we have seen in the previous section, given a C^*-algebra $\mathcal{A}$ and a representation π, an important role is plaid by the commutant $\pi(\mathcal{A})'$. The *double commutant* (or *bicommutant*) $\pi(\mathcal{A})'' \equiv (\pi(\mathcal{A})')'$ is also a useful mathematical concept. For example, if $\pi(\mathcal{A})$ is irreducible then $\pi(\mathcal{A})' = \{\lambda \mathbf{1}\,,\ \lambda \in \mathbf{C}\}$ and $\pi(\mathcal{A})'' = \mathcal{B}(\mathcal{H})$.

An interesting question is the relation between $\pi(\mathcal{A})''$ and $\pi(\mathcal{A})$. For this purpose, it is convenient to consider the strong and weak closures $\overline{\pi(\mathcal{A})}^s$ and $\overline{\pi(\mathcal{A})}^w$, respectively, with respect to the corresponding Hilbert topologies. As we shall see below, the two closures coincide and define a (concrete) C^*-algebra.

Definition 2.7.1 *A * subalgebra $\mathcal{A}$ of $\mathcal{B}(\mathcal{H})$ is said to be a* **Von Neumann algebra** *if it is equal to its bicommutant, $\mathcal{A} = \mathcal{A}''$.*

The following *Von Neumann bicommutant theorem* answers the questions raised above.

Theorem 2.7.2 *For a * subalgebra $\mathcal{A} \subseteq \mathcal{B}(\mathcal{H})$, with identity, the following three properties are equivalent*
i) $\mathcal{A} = \mathcal{A}''$,
ii) $\mathcal{A}$ is weakly closed,
iii) $\mathcal{A}$ is strongly closed.

Proof. We briefly sketch the proof.

i) $\Rightarrow$ ii). We show that quite generally the bicommutant is weakly closed. In fact, if $\mathcal{A}'' \ni A_n \overset{w}{\to} A$, then $\forall A' \in \mathcal{A}'$, $A_n A' = A' A_n$ and, $\forall x, y, \in \mathcal{H}$,

$$(x, (AA' - A'A)\,y) = (x, ((A - A_n)A' - A'(A_n - A))\,y)$$

can be made as small as we like, by choosing n sufficiently large. Hence $AA' = A'A$, i.e. $A \in \mathcal{A}''$.

ii) $\Rightarrow$ iii) is obvious since strong convergence implies weak convergence.

iii) $\Rightarrow$ i). We have to show that $\mathcal{A}$ is strongly dense in its bicommutant. For this purpose, we first show that $\forall x \in \mathcal{H}$, $\mathcal{A}'' x \subseteq \overline{\mathcal{A}\,x}^{\,s} \equiv \mathcal{H}_{\mathcal{A}x}$. In fact, if P is the projection on $\mathcal{H}_{\mathcal{A}x}$, then $P \in \mathcal{A}'$, since by definition, $\forall A \in \mathcal{A}$, $A\mathcal{H}_{\mathcal{A}x} \subseteq \mathcal{H}_{\mathcal{A}x} = P\mathcal{H}$, i.e. $\forall y \in \mathcal{H}, AP\,y \in P\mathcal{H}$, so that $PAP\,y = AP\,y$, which implies $PAP = AP$ and

$$PA = (A^*P)^* = (PA^*P)^* = PAP = AP.$$

Then, $\forall A'' \in \mathcal{A}''$, $A''\,x = A''Px = PA''x \in P\mathcal{H}$, $(Px = x$, because $\mathbf{1} \in \mathcal{A})$.

We can now prove that given $A'' \in \mathcal{A}''$ any of its strong neighborhoods

$$\mathcal{N}_{\epsilon,(x_1,\ldots,x_j)}(A'') \equiv \{A \in \mathcal{B}(\mathcal{H});\ \|(A - A'')x_k\| \le \epsilon,\ k = 1,\ldots,j\},$$

where $x_1, \ldots, x_j, \in \mathcal{H}$, contains an element of $\mathcal{A}$. In order to see this, given $x_1, \ldots, x_j$ one repeats the above argument for the j-fold direct sum $\widehat{\mathcal{H}}$ of $\mathcal{H}$; then, if $A'' \in \mathcal{A}''$, one has $\widehat{A''} = A'' \oplus \ldots \oplus A'' \in \hat{\mathcal{A}}''$ and $\hat{A}''\hat{x} \in \overline{\hat{\mathcal{A}}\hat{x}}$, where the elements $\hat{A} \in \hat{\mathcal{A}}$ are of the form $\hat{A} = A \oplus \ldots \oplus A$, $A \in \mathcal{A}$.

One of the consequences of the (Von Neumann bicommutant) theorem is that each representation π of a C^*-algebra $\mathcal{A}$ identifies a Von Neumann algebra $\overline{\pi(\mathcal{A})}^{\,w}$, which involves the weak Hilbert topology defined by the representation and is therefore a representation dependent concept.

Whereas the C^*-algebras can be regarded as the non-commutative analog of the algebras of continuous functions (on a compact space), the Von Neumann algebras can be regarded as the non-commutative analog of the bounded measurable functions, a concept which involves the measure, i.e. a positive linear functional on the algebra of continuous functions. Indeed, given a measure μ the bounded measurable functions are pointwise limits of continuous functions, almost everywhere with respect to μ, i.e. they belong to the weak closure of the algebra of continuous functions.

From an algebraic point of view Von Neumann algebras can be characterized as C^*-algebras which are the dual of a Banach space (also called W^* algebras). [17]

---

[17]S. Sakai, *C*-algebras and W*-algebras*, Springer 1971.

Chapter 3

The quantum particle

3.1 The Weyl algebra and the Heisenberg group

The classical (massive) particle is the simplest classical system and in classical mechanics its states are defined by the position q and the momentum p. In the more general framework discussed in Sect. 1.2, a classical particle is defined by the algebra of observables $\mathcal{A}$, which, in the case of compact phase space, can be obtained as the sup-norm closure of the polynomial algebra generated by q and p. Correspondingly, one is led to define the *quantum particle* by an algebra of observables "generated" by q and p,[1] which satisfy the Heisenberg commutation relations

$$[q\,,\,p] = i, \quad [q\,,\,q] = 0, \quad [p\,,\,p] = 0, \tag{3.1.1}$$

where for simplicity we have put $\hbar = 1$, by a suitable choice of units, and considered the case of one space dimension, the generalization being straightforward.

Here we meet the technical problem of giving a precise meaning to the word "generated". The vector space, over the field of complex numbers, generated by q and p, with Lie products given by eqs. (3.1.1), is a Lie algebra and it is called the *Heisenberg Lie Algebra*. In the Heisenberg formulation of Quantum Mechanics, the basic algebra is the (enveloping) algebra generated through sums and products by q and p, with commutation relations defined by the above Lie products, briefly called the *Heisenberg Algebra*. From a technical point of view, this is not the best choice and it does not fall into the general mathematical framework discussed in Sect. 1.3. In fact, the Heisenberg relations (3.1.1) imply that q and p cannot be self-adjoint

---

[1]Now we consider the case in which there is no other degree of freedom than q and p and no constraint on the position.

elements of a C^*-algebra, since they cannot be given a (finite) norm. In fact, eqs. (3.1.1) imply

$$[p, q^n] = -i\, n q^{n-1}$$

and therefore, if a C^* norm could be assigned to q and p,

$$n\, ||q^{n-1}|| \leq ||p\, q^n|| + ||q^n\, p|| \leq 2\, ||p||\, ||q||\, ||q^{n-1}||.$$

Now, $||q^{n-1}|| = ||q||^{n-1}$ cannot vanish, since otherwise $q = 0$ and eqs. (3.1.1) cannot hold. Hence, the above inequality implies

$$||q||\, ||p|| \geq n/2, \quad \forall n \in \mathbf{N},$$

i.e. $||q||, ||p||$ cannot both be finite.

The physical reason for this mathematical obstruction is that, strictly speaking, q and p are not observables in the operational sense discussed in Sect. 1.3; due to the scale bounds of experimental apparatuses, one actually measures only bounded functions of q and p, (namely the position inside the volume accessible by the experimental apparatus and the momentum inside an interval given by the energy bounds set by the apparatus). Thus, a formulation based on the Heisenberg algebra involves an (in fact physically harmless) extrapolation with respect to the operational definition of observables.

A solution of this merely technical difficulty was given by Weyl, who suggested to consider the polynomial algebra generated by the following bounded (formal) functions of q and p:

$$U(\alpha) = e^{i\alpha q}, \quad V(\beta) = e^{i\beta p}, \quad \alpha, \beta \in \mathbf{R}^s, \tag{3.1.2}$$

(in the case of s space dimensions), called the *Weyl operators*. [2]

The algebraic properties of the Weyl operators can be inferred by using their (formal) relations with the q's and p's.

1) *(Weyl commutation relations)* The Heisenberg relations (3.1.1) take the following form in terms of the Weyl operators

$$U(\alpha)\, V(\beta) = V(\beta)\, U(\alpha) e^{-i\alpha\beta}, \tag{3.1.3}$$

$$U(\alpha)U(\beta) = U(\alpha + \beta), \quad V(\alpha)V(\beta) = V(\alpha + \beta) \tag{3.1.4}$$

(commutation relations in Weyl form or, briefly, *Weyl relations*). This follows easily from the Baker-Hausdorff (BH) formula [3]

$$e^A e^B = e^{A+B+1/2[A,B]}, \tag{3.1.5}$$

---

[2]H. Weyl, *The Theory of Groups and Quantum Mechanics*, Dover 1931. Such operators can be interpreted as (limits of) observables associated to measurements of periodic functions of q, p.

[3]See e.g. N. Jacobson, *Lie Algebras*, Interscience 1962, Chap. 5, p. 170; R.M. Wilcox, Jour. Math. Phys. **8**, 962 (1967).

which holds if $C \equiv [A, B]$ commutes with A and B.

The BH formula can be easily derived under the general assumption that A, B and $A + B$ are densely defined operators in a Hilbert space with a common dense domain of analytic vectors (namely a domain on which one can freely apply $A, B, A + B$, and their exponentials). In fact, putting

$$G(\alpha) \equiv e^{-\alpha A} B e^{\alpha A},$$

one has

$$dG(\alpha)/d\alpha = -e^{-\alpha A}[A, B]e^{\alpha A} = -C,$$

which implies $G(\alpha) = -\alpha C + G(0)$, i.e.

$$B e^{-\alpha A} = e^{-\alpha A} B + \alpha C \, e^{-\alpha A}.$$

Then, putting

$$F(\alpha) \equiv e^{-\alpha B} e^{-\alpha A} e^{\alpha(A+B)} \exp(\alpha^2 C/2),$$

one has $dF(\alpha)/d\alpha = 0$, which implies

$$F(\alpha) = F(0) = 1,$$

i.e. the validity of the BH formula.

The abstract algebra $\mathcal{A}_W$ generated by (abstract) elements $U(\alpha), V(\beta)$, $\alpha, \beta \in \mathbf{R}$ (through complex linear combinations and products), satisfying eqs.(3.1.3), (3.1.4), is called the *Weyl algebra*. [4]

2) (*Unitarity*) There is a natural definition of $*$ on $U(\alpha), V(\beta)$, corresponding to the self-adjointness of q and $p_,$:

$$U(\alpha)^* = U(-\alpha), \quad V(\beta)^* = V(-\beta),$$

so that, by the Weyl relations, $U(\alpha), V(\beta)$ are unitary

$$U(\alpha)^* U(\alpha) = U(\alpha) U(\alpha)^* = \mathbf{1},$$

and similarly for $V(\beta)$.

3) (*Norm*) To get a C^*-algebra one has to assign a C^* norm to the elements of $\mathcal{A}_W$ and take the norm closure. The C^* condition requires

---

[4]It is worthwhile to remark that the Heisenberg algebra can be recovered from the Weyl algebra under the mild condition that the Weyl operators are represented by weakly continuous (Hilbert space) operators (see below). On the other side, the Weyl algebra can be constructed from the Heisenberg algebra under suitable regularity conditions, like that the common dense domain, on which the Heisenberg relations hold, is a domain of essential self-adjointness for $N \equiv q^2 + p^2$ (for a discussion of these problems see e.g. C.R. Putnam, *Commutation Properties of Hilbert Space Operators and Related Topics*, Springer 1967). The advantage of the Weyl operators and the Weyl commutation relations is that they are not affected by domain problems and they maintain a meaning also when the Heisenberg relations cannot be derived.

that U, V and $U V$ have norm equal to one and therefore so must have any monomial of U's and V's, since by the Weyl relations it can be reduced to a product $U V$ (times a phase factor). One can show that also the norm of an arbitrary element of the Weyl system is fixed, in the sense that there is a unique C^* norm on $\mathcal{A}_W$ such that the completion of $\mathcal{A}_W$, with respect to it, is a C^* algebra.[5] Such an abstract C^*-algebra will still be denoted by $\mathcal{A}_W$ and called the *Weyl C^*-algebra.*

In conclusion, the quantum particle can be defined as the physical system characterized by the Weyl C^*-algebra. The first basic problem is then to determine the states of a quantum particle, i.e. the *representations of the Weyl C^*-algebra $\mathcal{A}_W$.*

The relation between the Heisenberg Lie algebra and the Weyl algebra involves the standard mathematical step of associating a Lie group to a given Lie algebra through the exponential map: in our case the group is the *Heisenberg Lie group.* It can be defined as the Lie group with elements labeled by triples (α, β, λ), $\alpha, \beta \in \mathbf{R}^s, \lambda \in \mathbf{R}$, with the following group law $(\alpha, \beta, \lambda)(\alpha', \beta', \lambda') = (\alpha + \alpha', \beta + \beta', \lambda + \lambda' + [\alpha'\beta - \alpha\beta']/2)$. Such a composition law follows easily from the identification: $(\alpha, \beta, \lambda) \leftrightarrow \exp i(\alpha q + \beta p + \lambda \mathbf{1})$ and the Weyl relations. It is a non-compact Lie group. The mathematical description of the states of a quantum particle therefore amounts to find the representations of the Weyl C^*-algebra, or, equivalently, of the unitary representations of the Heisenberg group. [6]

3.2 Von Neumann uniqueness theorem

The classification of the representations of the Weyl algebra is trivialized by Von Neumann theorem, according to which all the regular irreducible representations are unitarily equivalent. The regularity condition is an extremely mild request and it is standard in the theory of representations of Lie algebras and Lie groups.

Definition 3.2.1 *A representation π of the Weyl algebra (or a unitary representation of the Heisenberg group) in a separable Hilbert space $\mathcal{H}$ is* **regular** *if $\pi(U(\alpha))$, $\pi(V(\beta))$ are (one-parameter groups of unitary operators) strongly continuous in α, β, respectively.*

For unitary operators in a Hilbert space, *strong continuity* is actually equivalent to the apparently weaker condition of *weak continuity*, since, for unitary operators $U(t)$, one has

$$||(U(t) - \mathbf{1}) \Psi||^2 = 2 ||\Psi||^2 - 2 \, Re \, (\Psi, U(t) \, \Psi).$$

---

[5] J. Slawny, Comm. Math. Phys. **24**, 151 (1971); J. Manuceau, M. Sirugue, D. Testard and A. Verbeure, Comm. Math. Phys. **32**, 231 (1973).

[6] See e.g. G.B. Folland, *Harmonic Analysis in Phase Space*, Princeton Univ. Press 1989.

For separable spaces, weak continuity is in turn equivalent to weak measurability, [7] i.e. to the property that the matrix elements $(\Phi, U(t)\Psi)$ are (Lebesgue) measurable functions. This shows how reasonable is the regularity condition. Moreover, by Stone's theorem on one-parameter groups of strongly continuous unitary operators, [8] regularity is equivalent to the existence of the generators of the Weyl operators, namely q and p, as self-adjoint unbounded operators in $\mathcal{H}$. Furthermore, one can show that they have a common invariant dense domain.[9] Thus, regularity allows to reconstruct the Heisenberg algebra from the Weyl algebra.

Theorem 3.2.2 *(Von Neumann) All the regular irreducible representations of the Weyl C^*-algebra are unitarily equivalent.*

Proof. It is convenient to introduce the following more general Weyl operators

$$W(\alpha, \beta) \equiv e^{-i\alpha\beta/2}V(\beta)\,U(\alpha) = e^{i\alpha\beta/2}U(\alpha)\,V(\beta), \tag{3.2.1}$$

which have the following properties:

$$W(\alpha, \beta)^* = W(-\alpha, -\beta), \quad W(\alpha, \beta)^*\,W(\alpha, \beta) = \mathbf{1}, \tag{3.2.2}$$

$$W(\alpha, \beta)W(\gamma, \delta) = W(\alpha + \gamma, \beta + \delta)e^{-i(\alpha\delta - \gamma\beta)/2}$$
$$= e^{i(\gamma\beta - \alpha\delta)}\,W(\gamma, \delta)\,W(\alpha, \beta). \tag{3.2.3}$$

The idea of the proof is to show that for any regular irreducible representation π in a Hilbert space $\mathcal{H}$, there exists a vector $\Psi_0 \in \mathcal{H}$, called the *Fock state* vector, which is cyclic by the irreducibility of π, with the property that

$$(\Psi_0, \pi(W(\alpha, \beta))\,\Psi_0) = \exp\left[-(|\alpha|^2 + |\beta|^2)/4\right]. \tag{3.2.4}$$

Since monomials of W's can be reduced to a single W, apart from a phase factor, eq. (3.2.4) determines all the expectations of $\mathcal{A}_W$ on Ψ_0 and by iii) of the GNS representation theorem π is unitarily equivalent to the GNS representation of the state ω_F defined by

$$\omega_F(W(\alpha, \beta)) = \exp\left[-(|\alpha|^2 + |\beta|^2)/4\right].$$

Such a representation is called the *Fock representation*.

---

[7]J. Von Neumann, Ann. Math.(2), **33**, 567 (1932). For a simple account see M. Reed and B. Simon, *Methods of Modern Mathematical Physics*, Vol. I, Academic Press 1972, Chap. VIII, Sect. 4.

[8]See e.g. Appendix F below.

[9]O. Bratteli and D.W. Robinson, *Operator Algebras and Quantum Statistical Mechanics*, Vol. II, Springer 1981, Sect. 5.2.3.

To construct Ψ_0, we note that, thanks to the regularity of π, the integral

$$P \equiv (1/2\pi) \int d\alpha d\beta \, \exp\left[-(|\alpha|^2 + |\beta|^2)/4\right] \pi(W(\alpha, \beta)) = P^*$$

exists as a strong limit of Riemann sums, since $\exp\left[-(|\alpha|^2 + |\beta|^2)/4\right] \in L^1(\mathbf{R}^2)$ and $\pi(W(\alpha, \beta))$ is strongly continuous and bounded in α, β. Moreover, P cannot vanish, since, otherwise, $\forall \gamma, \delta \in \mathbf{R}$

$$(1/2\pi) \int d\alpha \, d\beta \, [e^{-(|\alpha|^2 + |\beta|^2)/4} \pi(W(\alpha, \beta))] e^{i(\gamma\beta - \alpha\delta)}$$

$$= \pi(W(-\gamma, -\delta)) P \pi(W(\gamma, \delta)) = 0,$$

i.e. the Fourier transform of the matrix elements of the operator in square brackets would vanish and therefore $\pi(W) = 0$. Finally,

$$P \, \pi(W(\alpha, \beta)) \, P = e^{-(|\alpha|^2 + |\beta|^2)/4} P, \qquad (3.2.5)$$

since the left hand side is

$$(2\pi)^{-2} \int d\gamma \, d\delta \, d\gamma' \, d\delta' \, e^{-(|\gamma|^2 + |\delta|^2 + |\gamma'|^2 + |\delta'|^2)/4} e^{-i(\alpha\delta' - \beta\gamma')/2}$$

$$\times e^{-i(\gamma(\beta+\delta') - \delta(\alpha+\gamma'))/2} \, \pi(W(\alpha + \gamma + \gamma', \beta + \delta + \delta'))$$

and by a change of variables, $\gamma + \gamma' = k - \alpha$, $\delta + \delta' = \nu - \beta$, $\gamma - \gamma' = \mu$, $\delta - \delta' = \lambda$, the integration is reduced to Gaussian integrals, which yield the r.h.s. of eq. (3.2.5). This equation implies $P^2 = P$, (by putting $\alpha = 0 = \beta$), so that P is a (non-zero) projection. Thus, $\exists \Psi_0$ with $P\Psi_0 = \Psi_0, \|\Psi_0\| = 1$ and by eq. (3.2.5) it satisfies eq. (3.2.4).

Actually, by the irreducibility of π, P is a one dimensional projection. In fact, if Φ is orthogonal to Ψ_0 and $P\Phi = \Phi$ one has

$$(\Phi, \pi(W(\alpha, \beta)) \, \Psi_0) = (P\Phi, \pi(W(\alpha, \beta)) P \, \Psi_0) = (\Phi, P\pi(W(\alpha, \beta)) P \, \Psi_0)$$

$$= \exp\left[-(|\alpha|^2 + |\beta|^2)/4\right] (\Phi, P \, \Psi_0) = 0.$$

This implies $(\Phi, A \, \Psi_0) = 0, \forall A \in \mathcal{A}_W$, and, since every vector of an irreducible representation is cyclic, one gets $\Phi = 0$.

Equation (3.2.4) is equivalent to $(-i\partial/\partial\alpha + \partial/\partial\beta + \beta)\omega_F(U(\alpha) \, V(\beta)) = 0$, i.e. $(q + ip)\Psi_0 = 0$.

3.3   The Schroedinger representation and wave function

Since by Von Neumann theorem all regular irreducible representations of the Weyl C^*-algebra are unitarily equivalent, it suffices to find one. Thus, we write down the so-called *Schroedinger representation* π_S. Henceforth, for simplicity $\pi_S(A)$, $A \in \mathcal{A}_W$ will be denoted by A. The Hilbert space is

$$\mathcal{H} = L^2(\mathbf{R}^s, d^s x), \quad s = \text{space dimension} \tag{3.3.1}$$

and $\forall \psi \in \mathcal{H}$ we put

$$(U(\alpha)\psi)(x) = e^{i\alpha x}\, \psi(x), \quad (V(\beta)\,\psi)(x) = \psi(x + \beta) \equiv \psi_\beta(x). \tag{3.3.2}$$

It is clear that $U(\alpha)$, $V(\beta)$ are unitary operators in $\mathcal{H}$ and that they define a strongly continuous representation of the Weyl C^*-algebra:

$$(U(\alpha)\,V(\beta)\psi)(x) = (U(\alpha)\,\psi_\beta)(x) = e^{i\alpha x}\, \psi(x + \beta),$$

$$(V(\beta)\,U(\alpha)\psi)(x) = e^{i\alpha(x+\beta)}\, \psi(x + \beta),$$

$$U(\alpha)U(\beta)\psi = U(\alpha + \beta)\psi, \quad V(\alpha)V(\beta)\psi = V(\alpha + \beta)\psi.$$

It remains to prove that the Schroedinger representation is irreducible. In fact, if not, there would exist an invariant non trivial (proper) subspace $\mathcal{H}_1$ and a vector $\phi \in \mathcal{H}_1^\perp$. Then, given $\psi \in \mathcal{H}_1$

$$0 = (\phi, U(\alpha)\,V(\beta)\psi) = \int dx e^{i\alpha x}\, \bar{\phi}(x)\psi_\beta(x),$$

i.e. the Fourier transform of $\bar{\phi}\psi_\beta$ vanishes and therefore so does $\bar{\phi}\psi_\beta$. Hence,

$$\text{supp}\phi \cap \text{supp}\psi_\beta = \emptyset, \quad \forall \beta \in \mathbf{R}$$

and since by varying β one can translate the support of ψ as one likes, ϕ must vanish.

The physical consequences displayed by the Schroedinger representation are that the pure states of a quantum particle are represented by vectors in $L^2(\mathbf{R}^s)$, i.e. by L^2-functions. They are called *Schroedinger wave functions*.

By Stone's theorem, the strong continuity of $U(\alpha), V(\beta)$ implies that the derivatives $-idU(\alpha)/d\alpha|_{\alpha=0}, -idV(\beta)/d\beta|_{\beta=0}$, exist as strong limits on a dense domain and define the corresponding generators q and p, respectively, as unbounded (densely defined) operators in L^2:

$$(q\psi)(x) = x\psi(x), \quad \psi \in D_q = \{\psi \in L^2, \ x\psi \in L^2\}, \tag{3.3.3}$$

$$(p\psi)(x) = -i(d/dx)\psi(x), \quad \psi \in D_p = \{\psi \in L^2, \ d\psi/dx \in L^2\}, \tag{3.3.4}$$

(here and in the following the derivative of an L^2 function is always understood in the sense of distributions).

The dense domain $D = \{\psi \in \mathcal{S}(\mathbf{R}^s)\}$ ($\mathcal{S}$ denotes the Schwartz space of C^∞ functions of fast decrease) is a common dense domain of essential self-adjointness [10] for q and p and it is easy to check that on D they satisfy the Heisenberg relations. Thus, one has also a representation of the Heisenberg algebra. Here, we realize a substantial difference between the classical and the quantum case: the position is represented by a multiplication operator and the momentum by a differential operator.

An observable is represented by a bounded (self-adjoint) operator A in $\mathcal{H}$ and the expectations $\omega_\psi(A)$, $\psi \in \mathcal{H}$, are given by

$$\omega_\psi(A) = \int dx\, \bar{\psi}(x)(A\psi)(x).$$

Explicitly, the action of A on ψ is specified by saying which function of q and p, A corresponds to. If A is a (bounded) function of q, say $F(q)$, then A is represented by the multiplication operator $F(x)$. The potential energy defined by the potential $V = V(q)$ is therefore represented by $V(x)$.

For such operators, the probabilistic interpretation is particularly simple in the Schroedinger representation, since

$$\omega_\psi(A(q)) = \int dx\, \bar{\psi}(x)A(x)\psi(x),$$

so that A can be regarded as a random variable with a probability distribution function $|\psi(x)|^2\, dx$. The situation is different for operators which are functions of the momentum (e.g. the kinetic energy $T = p^2/2m$ represented by $-\Delta/2m$) ; they cannot be interpreted as random variables with the same joint probability distribution $|\psi(x)|^2\, dx$. In fact, one has

$$\omega_\psi(p) = \int dx\, \bar{\psi}(x)(-id/dx)\psi(x),$$

which is not of the above form for $A(q)$.

This means that the description of a quantum system cannot be done in terms of the classical theory of probability; rather it provides the prototype of what is called quantum probability (see Sect. 2.4). On the other hand, since the Fourier transform is a one to one mapping of L^2 into L^2, a state vector of $\mathcal{H}$, described by $\psi(x)$, is equally well uniquely described by the

---

[10]For the basic concepts of self-adjointness and of essential self-adjointness see e.g. M. Reed and B. Simon, *Methods of Modern Mathematical Physics*, Vol. I (Functional Analysis), Academic Press 1972, Chap. VIII, Sect. 2. For the convenience of the reader a brief account, also in connection with the problem of the existence of the dynamics, is presented in Appendix F.

Fourier transform $\tilde{\psi}(k)$, so that the above expectation values can also be written in the following form

$$\omega_\psi(A(q)) = \int dk \bar{\tilde{\psi}}(k) A(id/dk) \tilde{\psi}(k),$$

$$\omega_\psi(p) = \int dk \bar{\tilde{\psi}}(k) \, k \, \tilde{\psi}(k) = \int dk |\tilde{\psi}|^2(k) \, k.$$

In this form the probabilistic interpretation is simple for the functions of the momentum, but not for the functions of the position.

Since any $\psi(x) \in L^2$ completely defines a state, it determines the expectations and therefore the probabilistic interpretations for both q and p. In fact, it provides much richer information than the probability distribution for q, namely $|\psi(x)|^2$, since the multiplication by a phase factor: $\psi(x) \to \exp(ip_0 x) \psi(x)$ does not change $|\psi(x)|^2$, but it changes the Fourier transform of ψ and therefore the probability distribution for p. As a consequence of the non abelianess of the algebra generated by q and p, the probability distributions for the two variables cannot be prescribed in a completely independent way, as in the classical case, since the Heisenberg uncertainty relations must hold.

A typical phenomenon of the quantum mechanical description is that, given two states represented by ψ_1, $\psi_2 \in L^2$, also

$$\psi(x) = \psi_1(x) + \psi_2(x)$$

is a state vector, which is said to be a *superposition* of the state vectors ψ_1, ψ_2. It is worthwhile to remark that the probability distribution of the observable $F(q)$, defined by $\psi(x)$, is *not* the sum of the probability distributions $|\psi_1(x)|^2$ and $|\psi_2(x)|^2$, but it contains mixed or *interference* terms $\bar{\psi}_1(x)\psi_2(x) + \bar{\psi}_2(x)\psi_1(x)$. This is the origin of the *wave properties* of a quantum particle, since it is responsible for constructive or destructive interference phenomena as for optical waves.

3.4   Gaussian states. Minimal Heisenberg uncertainty

From the discussion of the previous section it follows easily that the state ω represented by the following Gaussian wave function

$$\psi(x) = (2\pi a^2)^{-1/4} \exp\left[-(x-x_0)^2/4a^2\right] e^{ibx} \tag{3.4.1}$$

describes a quantum particle "essentially" localized in x_0, with $\Delta_\omega q = a$ and $\omega(p) = b$. Briefly one also says that the above wave function describes a *wave packet* peaked at $x = x_0$, with a spread $\Delta x = a$. State vectors of this form have the characteristic property of minimizing the Heisenberg uncertainty, namely

$$(\Delta_\omega q)\,(\Delta_\omega p) = 1/2. \qquad (3.4.2)$$

This can be checked by an easy computation of Gaussian integrals, since the Fourier transform of a Gaussian is a Gaussian. It is not difficult to see that the above equation selects the Gaussian state vectors. In fact, putting

$$\tilde{q} \equiv q - \omega(q), \quad \tilde{p} \equiv p - \omega(p)$$

one has

$$1 = |\omega([\tilde{q}, \tilde{p}])| = 2\,|\mathrm{Im}\,\omega(\tilde{q}\tilde{p})| \leq 2\,|\omega(\tilde{q}\tilde{p})|$$
$$\leq 2\,\Delta_\omega(q)\,\Delta_\omega(p),$$

where the Cauchy-Schwarz' inequality has been used in the last step. The minimal uncertainty is obtained iff both inequalities above become equalities i.e. if $\omega((q - \omega(q))(p - \omega(p)))$ is purely imaginary and

$$(q - \omega(q))\psi_\omega = \mu(p - \omega(p))\psi_\omega, \quad \mu \in \mathbf{C}.$$

Thus, $\mu = -i\lambda$, $\lambda \in \mathbf{R}$ and one has the equation, (ψ in the domain of the operators involved),

$$(x + \lambda(d/dx) - <x> - i\lambda <p>)\psi = 0.$$

Putting

$$\psi(x) = \exp\left[-(x - <x>)^2/2\lambda\right] e^{i<p>(x-<x>)}\phi(x),$$

one easily finds that ϕ is a (normalization) constant, $\lambda = 2\,\Delta_\omega(q)^2$ and the above Gaussian wave function is recovered (the above mentioned domain conditions are obviously satisfied).

The above Gaussian wave functions describe states which are the closest ones to the classical states since they describe a particle with the best localization in both position and momentum (respectively around $<x> = x_0$ and $<p>$).

The Gaussian state vector with $<x> = 0$, $<p> = 0$, $\Delta x = 1/\sqrt{2}$ is the representative of the Fock state ω_F, introduced in Sect.3.2, as one can easily check by a Gaussian integration. It follows easily that the Fock state vector can be characterized by the following equation

$$(q + ip)\Psi_0 = 0. \qquad (3.4.3)$$

It may be worthwhile to mention that a position uncertainty of 10^{-8} cm (the atomic radius) allows a momentum uncertainty up to $\simeq 0.5 \cdot 10^{-27}$ g cm sec^{-1}, which becomes critical only for particles of very small size, like the electron ($m_e \simeq 10^{-27}$ g).

Chapter 4

Quantum dynamics. The Schroedinger equation

4.1 Quantum dynamics. The quantum Hamiltonian

The analysis of the mathematical description of a physical system in terms of observables and states on the basis of measurements performed on the system can be extended to the relation between measurements at different times. In this way one gets a general mathematical description of a dynamical system in algebraic terms. [1]

If A is an observable defined by an experimental apparatus at a given time e.g. $t = 0$, the same kind of measurement performed at a later time t defines the corresponding observable A_t. Due to the intrinsic limitation of the precision of experimental measurements, one cannot make a measurement at a sharp time, a finite time interval being always involved. Hence, some time averaging is always involved and the label t can be given the meaning of the central value of the non-zero time interval during which the measurement is done (and over which a time average takes place).

In the following discussion, for simplicity, we restrict our considerations to non-dissipative systems so that, by choosing experimental apparatuses whose functioning does not depend on time, the relation between measurements on the system at different times, say t_1, t_2, only depends on the time difference $t_2 - t_1$. Thus, the discussion of Sect. 1.3 indicates that the algebra $\mathcal{A}$ generated by the observables is the same at any time and that the time

---

[1]For a general approach to quantum dynamics and a general discussion of C^* dynamical systems see O. Bratteli and D.K. Robinson, *Operator Algebras and Quantum Statistical Mechanics*, Vol. I, Springer 1987, Sects. 2.7.1, 3.2.2.

translation $A \to A_t \equiv \alpha_t(A)$ preserves all the algebraic properties, including the $*$, i.e. it defines a $*$- automorphism α_t of $\mathcal{A}$. When t varies one actually gets a one-parameter group of automorphisms $\alpha_t, t \in \mathbf{R}$, (*algebraic dynamics*), since clearly $\alpha_{t_1}(\alpha_{t_2}(A)) = \alpha_{t_1+t_2}(A)$.

The stability of the algebra of observables under time evolution is therefore a necessary physically motivated requirement. In a constructive approach, the identification of the algebra of observables may not be trivial and therefore, if one starts with a putative candidate $\mathcal{A}$ for the algebra of observables, it is possible that one must enlarge $\mathcal{A}$ to obtain stability under time evolution. [2] By the same physical considerations, the family of physical states must be stable under time evolution, $\alpha_t{}^*, t \in \mathbf{R}$.

The general considerations of Sect. 1.3 and the unavoidable time average involved in any actual measurement lead to the condition that, for any state ω, the expectation $\omega(\alpha_t(A))$ is continuous in t, technically that α_t is *weakly continuous* (see the similar discussion in Sect. 1.2 for the classical case).

In conclusion, the general mathematical description of a physical dynamical system involves a triple $(\mathcal{A}, \mathcal{S}, \alpha_t, t \in \mathbf{R})$, where $\mathcal{A}$ is the C^*-algebra of observables, $\mathcal{S}$ the set of physical states and α_t the algebraic dynamics.

Since time translation is a "physically realizable operation", it is natural to associate to a given (physical) state ω the set of states which are either related to it by action of observables or by time translations. Thus, one is led to consider representations π *stable under time translations*, in the sense that, if ω belongs to the folium of π, so does $\alpha_t^*\omega$.

For an irreducible representation π, this implies that $\pi \circ \alpha_t$ is unitarily equivalent to π and therefore α_t is implemented by a unitary operator $U(t)$:

$$\pi(\alpha_t(A)) = U(t)^{-1} \pi(A) U(t), \quad \forall A \in \mathcal{A} \ . \tag{4.1.1}$$

The weak continuity of α_t then implies that $U(t)$ may be chosen as a weakly continuous group (in t). [3] Then, by Stone's theorem,[4] one can write

$$U(t) = \exp(-i\,t\,H), \quad \forall t \in \mathbf{R}, \tag{4.1.2}$$

where the generator H is a self-adjoint operator in $\mathcal{H}_\pi$ with dense domain D_H (of self-adjointness) and is the strong derivative of $U(t)$ on D_H, i.e.

$$\text{strong-}\lim_{t \to 0} i\, t^{-1}(U(t) - \mathbf{1}) \Psi = H \Psi, \quad \forall \Psi \in D_H.$$

Furthermore, D_H contains a dense domain D invariant under $U(t)$, on which H is essentially self-adjoint (see Appendix F). Then, $\forall \Psi \in D$, putting

[2] M. Fannes and A. Verbeure, Comm. Math. Phys. **35**, 257 (1974); G.L. Sewell, Lett. Math. Phys. **6**, 209 (1982); G. Morchio and F. Strocchi, Jour. Math. Phys. **28**, 622 (1987).

[3] O. Bratteli and D.K. Robinson, loc. cit., Vol. I, Ex. 3.2.14, Ex. 3.2.35.

[4] See M. Reed and B. Simon, *Methods of Modern Mathematical Physics*, Vol. I, Academic Press 1972, Sect. VIII.4; for a brief sketch see Appendix F below.

$\Psi(t) \equiv U(t)\Psi$, we have

$$i\frac{d}{dt}\Psi(t) = H\,\Psi(t). \tag{4.1.3}$$

This is the (time) *evolution equation in Schroedinger form* and H is called the (quantum) *Hamiltonian*.[5]

From a constructive point of view, a quantum mechanical model is defined by specifying the Hamiltonian H as a self-adjoint operator in the representation space $\mathcal{H}_\pi$ and the evolution problem amounts to solve eq. (4.1.3) for any initial data Ψ in an invariant domain D of essential self-adjointness of H. This defines the "propagator" $U(t)$, $t \in \mathbf{R}$, for any $\Psi \in \mathcal{H}$. Actually, by Stone theorem, the existence and uniqueness of solutions of eq. (4.1.3), i.e. of $U(t)$, is equivalent to the self-adjointness of H.

The evolution problem can be equivalently formulated in terms of the evolution of the observables, rather than of the states. From eqs. (4.1.1), (4.1.2), omitting the symbol π and putting $A(t) \equiv U(t)^* A U(t)$, one has

$$\frac{d}{dt}A(t) = i[H, A(t)], \quad \forall A \in \mathcal{A}\ . \tag{4.1.4}$$

These are the *evolution equations in Heisenberg form* and it is clearly enough to solve them for a set of A's which generate the entire algebra.[6] Such solution defines a one parameter unitary group $U(t), t \in \mathbf{R}$, such that $\alpha_t(A) = U(t)^* A U(t)$, $\forall A \in \mathcal{A}$.

The above considerations apply to any quantum system, including those with infinite degrees of freedom. For the simple case of a quantum particle, the Weyl algebra is generated by the q, p, so that their time evolution determines that of the Weyl algebra. The evolution equations for q, p read

$$dq/dt = i[H, q], \quad dp/dt = i[H, p].$$

They are the strict analog of the classical Hamilton equations with the Poisson brackets replaced by the commutators. Indeed, if e.g. $H = T(p) + V(q)$, (with T and V regular functions), from the Heisenberg commutation relations one gets [7]

$$dq/dt = \partial H/\partial p, \quad dp/dt = -\partial H/\partial q.$$

---

[5] Here we keep working in suitable units so that $\hbar = 1$; otherwise one would have $U(t) = \exp(-itH/\hbar)$ and an $\hbar$ would appear in front of d/dt in eq. (4.1.3).

[6] The above equation does not have a purely algebraic content, since in general the Hamiltonian is an unbounded operator and therefore it does not belong to the C^*-algebra of observables; what one can actually measure are bounded functions of the Hamiltonian (see the analogous discussion in connection with the Heisenberg and the Weyl algebras in Sect. 3.1). Thus, to give a meaning to the derivative $d\alpha_t(A)/dt$, one must work in a given representation. In fact, for general quantum systems (including those with infinite degrees of freedom, for which Von Neumann theorem does not apply and more than one representation is available) the Hamiltonian is a representation dependent concept.

[7] This may easily checked in the Schroedinger representation, where q and p correspond to differential operators.

4.2   The dynamics of a free quantum particle

In Sect. 3.3 we have characterized the mathematical description of a quantum particle in terms of the Schroedinger representation; therefore, to define the dynamics, one has to specify the Hamiltonian as an operator in $\mathcal{H} = L^2(\mathbf{R}^s)$.

For this purpose, we first consider the case of a free quantum particle of mass m, which, in analogy with the classical case, is defined by the property that the momentum is a constant of motion and the position is a linear function of time:

$$dp(t)/dt = 0, \quad dq(t)/dt = p/m.$$

It is easy to recognize the Hamiltonian $H = H(q, p)$, which yields the above equations as evolution equations in Heisenberg form. In fact, from eqs. (4.1.4) we have $0 = dp(t)/dt = i\,[H, p(t)] = i\,[H, p]$, so that H must be a function of p only. Furthermore, since $i\,[H, q] = p/m$, $\tilde{H} \equiv H - p^2/2m$ commutes with both p and q and therefore with the whole Weyl algebra (generated by them). Since the Schroedinger representation is irreducible, $\tilde{H}$ must be a multiple of the identity and since, as in the classical case, the Hamiltonian is defined up to a constant multiple of the identity, without loss of generality, one can take

$$H = H_0 \equiv \frac{p^2}{2m} = -\frac{\Delta}{2m}, \tag{4.2.1}$$

where the explicit representation of p as a differential operator has been used. Hence, the evolution equation in Schroedinger form reads

$$i\frac{\partial}{\partial t}\psi(x, t) = -\frac{\Delta}{2m}\psi(x, t). \tag{4.2.2}$$

This is the *Schroedinger equation for a free quantum particle* in $\mathbf{R}^s$; for any initial data $\psi(x) \in D(H_0) \equiv \{\psi \in L^2, \Delta\psi \in L^2\}$, the domain of self-adjointness of H_0 (see Appendix F), the solution exists and it gives the wave function $\psi(x, t)$ at time t. In fact, the equation is easily solved by Fourier transform in the variable x:

$$\tilde{\psi}(k, t) = e^{-it\,k^2/2m}\tilde{\psi}(k, 0), \tag{4.2.3}$$

so that

$$\psi(x, t) = (2\pi)^{-s/2}\int d^s k\, e^{ikx}\, e^{-itk^2/2m}\, \tilde{\psi}(k, 0) \equiv \int dx'\, G_0(x, x'; t)\,\psi(x').$$

The function $G_0(x, x'; t)$ is called the kernel of $\exp(-iH_0 t)$ as integral operator in L^2.

It is interesting to work out the time evolution of the Gaussian wave functions describing states close to the classical ones. For simplicity we consider the one dimensional case; the wave function

$$\psi(x,0) = C\,e^{-(x/2\Delta_0 x)^2}\,e^{imvx/\hbar},$$

where C is a normalization constant and the physical constant $\hbar$ has been reestablished for the numerical calculations, describes a quantum particle localized at the origin with average momentum mv. At time t we have

$$\psi(x,t) = C\left(1 + \frac{2i\hbar t}{m(2\Delta_0 x)^2}\right)^{-1/2} e^{imv^2 t/2\hbar}\,e^{imvx/\hbar}$$

$$\times \exp\left[-\frac{(x-vt)^2}{(2\Delta_t x)^2}\left(1 + \frac{im(2\Delta_0 x)^2}{2\hbar t}\right)\right],$$

where

$$(2\Delta_t x)^2 \equiv (2\Delta_0 x)^2\left[1 + \left(\frac{2\hbar t}{m(2\Delta_0 x)^2}\right)^2\right],$$

so that

$$|\psi(x,t)|^2 = |C|^2(\Delta_0 x/\Delta_t x)^2\,e^{-2(x-vt)^2/(2\Delta_t x)^2}.$$

This shows that the peak of the wave packet moves with velocity $v = <p>/m$, as for the motion of a free classical particle, and the spread $\Delta_t x$ increases with time (*spreading of the wave packet in time*).[8]

It is important to stress that the time t needed to double the initial spread crucially depends on the mass of the particle: for an atomic particle like a hydrogen atom, $m = 1.7 \times 10^{-24}$ gr, $\Delta_0 x = 10^{-8}$ cm, one has $t \simeq 6 \times 10^{-13}$ sec, whereas for a macroscopic particle of mass $m = 10^{-3}$ gr, with $\Delta_0 x = 10^{-3}$ cm, one has $t \simeq 0.3 \times 10^{11}$ years.

4.3 Quantum particle in a potential

In classical mechanics the dynamics of a particle in a (position dependent) potential $V(q)$ is governed by the Hamiltonian $H = p^2/2m + V(q)$. In close analogy, the quantum particle in a potential is described by a Hamiltonian of the same form (with a proper interpretation of the symbols) and in the Schroedinger representation it reads

$$H = -\frac{\Delta}{2m} + V(x). \tag{4.3.1}$$

---

[8]The increase of spreading depends on the condition that at $t = 0$ the wave function minimizes the Heisenberg uncertainty and it holds also for backward (in time) evolution.

To formulate the problem correctly, one should remark that, in general, the r.h.s. of the above equation does not uniquely identify a densely defined self-adjoint operator in L^2 and therefore, according to Stone's theorem, it does not (uniquely) define a dynamics. Even if $-\Delta/2m$ and $V(x)$ separately define self-adjoint operators, it is not guaranteed that their sum does; in general the above equation defines a symmetric operator, but symmetric operators may have several or no self-adjoint extensions.[9] This is a typical problem of functional analysis, which is crucial for the foundations of quantum dynamics also from a physical point of view, since different self-adjoint extensions give rise to different spectra and to different quantum dynamics.[10] It should also be stressed that the solution of this problem corresponds to the solution of the Cauchy problem for the Schroedinger equation, because the definition of H as a self-adjoint operator guarantees the existence of the unitary operator e^{-itH}, i.e. the existence of the time evolution for any initial data in L^2.

Simple, but powerful criteria for self-adjointness were given by Rellich and especially by Kato.[11] We list the basic results; for their discussion and proofs we refer the reader to the beautiful exposition in Reed and Simon book (Vol. II). A brief sketch is given in Appendix F.

Definition 4.3.1 *Let H_0, H_1 be densely defined operators in a Hilbert space $\mathcal{H}$, with domain $D(H_0) \subset D(H_1)$, then H_1 is said to be* **smaller** *in the sense of Kato if there exist real numbers $a < 1$ and b such that $\forall \Phi \in D(H_0)$*

$$||H_1 \Phi|| \le a\,||H_0 \Phi|| + b\,||\Phi||. \tag{4.3.2}$$

Theorem 4.3.2 *If H_0 is self-adjoint on $D(H_0)$ and H_1 is a hermitian operator smaller than H_0 in the sense of Kato, then $H = H_0 + H_1$, with domain $D(H_0)$ is self-adjoint.*

Theorem 4.3.3 *Let $V \in L^2(\mathbf{R}^3) + L^\infty(\mathbf{R}^3)$ be a real potential, then V is smaller than $H_0 = -\Delta/2m$ on the domain $D(H_0)$ of self-adjointness of H_0 and $-\Delta/2m + V(x)$ is self-adjoint on $D(H_0)$.*

---

[9] For a review of this problem, which appears in the theory of unbounded (Hilbert space) operators, see e.g. N.I. Akhiezer and I.M. Glazman, *Theory of Linear Operators in Hilbert Spaces*, Pitman 1981; M. Reed and B. Simon, *Methods of Modern Mathematical Physics*, Vol. II (Fourier Analysis, Self-Adjointness), Academic Press 1975, Chap. X.

[10] This problem is usually overlooked in the textbooks on Quantum Mechanics. A simple and clear discussion (also in connection with its physical relevance) is given in A.S. Wightman, Introduction to some aspects of the relativistic dynamics of quantum fields, in *High Energy Electromagnetic Interactions and Field Theory*, M. Lévy ed., Cargèse Lectures, Gordon and Breach 1967, pp. 171-291, Sect. 8. The problem is discussed at length in the above quoted book by M. Reed and B. Simon, Chap. X. Quite generally, this is one of the basic problems of the mathematical theory of Schroedinger operators.

[11] T. Kato, *Perturbation Theory for Linear Operators*, Springer 1966.

Theorem 4.3.4 *Let $V \in L^2_{loc}(\mathbf{R}^n)$, $V \geq 0$, then $H = -\Delta/2m + V$ is essentially self-adjoint on $C_0^\infty(\mathbf{R}^n)$, (the set of infinitely differentiable functions of compact support).*

The relevance of the above theorems is that they cover the case of the Coulomb potential, which governs atomic physics, and most of the potentials used in nuclear physics. From these theorems it follows that in Quantum Mechanics the N-body problem, with a two-body potential of the class covered by Kato theorems, has a unique solution global in time, whereas the corresponding problem in Newtonian (classical) mechanics is not known to have solutions global in time.

In conclusion, for a potential V in Kato class, the Cauchy problem for the evolution partial differential equation

$$i\frac{\partial}{\partial t}\psi(x,t) = \left(-\frac{\Delta}{2m} + V(x)\right)\psi(x,t), \qquad (4.3.3)$$

$$\psi(x,0) = \psi \in D(-\Delta),$$

called the *Schroedinger equation*, is well posed and the corresponding initial value problem has a unique solution global in time.

4.4 Appendix F: Hamiltonian self-adjointness and dynamics

In the construction of the observables as operators in a Hilbert space $\mathcal{H}$, one usually starts with unbounded operators like q and p and one faces the problem of giving a meaning to (bounded) functions of the canonical variables, as self-adjoint operators in $\mathcal{H}$. The general solution of this problem (operator calculus) is given by the spectral theorem, which allows the construction of (continuous) functions of an operator A, provided A is defined as a self-adjoint (or at least normal) operator.[12]

Thus, e.g. given the hermitian (unbounded) operator $p = -i \, d/dx$, in order to define $e^{i\alpha p}$, $\alpha \in \mathbf{R}$, one needs a self-adjoint extension of it. This problem is not a mere mathematical subtlety, but corresponds to concrete physical issues. For example, if one considers a quantum particle in a box, which for simplicity will be considered as the one dimensional interval $[0, 1]$, the unbounded operator $-i \, d/dx$ is naturally well defined as an hermitian operator on the $C^\infty([0, 1])$ functions of compact support contained in $(0, 1)$. This, however, is not enough for defining its spectrum and a corresponding operator calculus.

The physical reason is easily understood if one realizes that, by Stone theorem, the exponential $\exp i\alpha p$ is well defined iff p is self-adjoint; on the other hand, for α small $\exp i\alpha p$ acts as translation operator on $C_0^\infty([0, 1])$ so that to get it well defined on $L^2([0, 1], dx)$ one must specify what happens when the translated wave function starts hitting the boundary.

The one-parameter family of self-adjoint extensions of the differential operator $-i \, d/dx$, called p_θ, $0 \le \theta < 2\pi$, do in fact correspond to the possible behaviours at the boundary: p_θ is defined on the domain D_θ consisting of all absolutely continuous functions with square integrable derivatives, satisfying the boundary condition [13]

$$\psi(1) = e^{i\theta} \, \psi(0).$$

The complete orthonormal set $\{\psi_n(x) = \exp i\lambda_n x, \ n \in \mathbf{N}\}$ are eigenfunctions of p_θ with eigenvalues $\lambda_n = 2\pi n + \theta$ (this implies that p_θ is self-adjoint). None of the domains D_θ is stable under the action of the operator $q = x$, so that p_θ cannot be applied to $q \, \psi_n \notin D_\theta$ and this resolves the

---

[12] See e.g. the above quoted books by Akhiezer and Glazman and by Reed and Simon. A proof of the spectral theorem is given below.

[13] A function f is absolutely continuous in $[a, b]$ if $\forall \varepsilon$, $\exists \delta$ s.t. $\sum_i |f(x_i) - f(y_i)| < \varepsilon$, for any finite collection of disjoint intervals $[x_i, y_i]$ with $\sum_i |x_i - y_i| < \delta$. For p_θ see the book by M. Reed and B. Simon, quoted in footnote 9, Sect. X.1, Example 1.

paradox which would arise by naively taking the expectation of $[q, p] = i$ on ψ_n.[14]

In the case of a quantum particle living in $\mathbf{R}$, (more generally in $\mathbf{R}^s$), there is a unique self-adjoint extension of $-i\partial/\partial x$, characterized by the dense domain of square integrable functions with square integrable derivatives (no boundary condition is needed also on physical grounds).

The analysis of the self-adjoint extensions of differential operators becomes crucial for the construction of the evolution operator $\exp(-it\,H)$, $t \in \mathbf{R}$, starting from a formal Hamiltonian $H = -\Delta/2m + V(x)$. This problem is essentially solved by Kato's theorems mentioned in Sect. 4.3, (a beautiful account is given in Reed and Simon book, Vol.II. For the convenience of the reader, a sketchy account is given below).

We recall that, given a densely defined operator A on the Hilbert space $\mathcal{H}$, its *adjoint* A^* is so defined: its domain $D(A^*)$ is the set of all $x \in \mathcal{H}$, with the property that there exists x^* (depending on x) such that $\forall y \in D(A)$

$$(Ay, x) = (y, x^*),$$

(uniqueness of x^*, for given x, follows from $D(A)$ being dense), and on such a domain A^* is defined by $A^*x \equiv x^*$. Then, one has

$$(Ay, x) = (y, A^*x), \quad \forall\, y \in D(A), \ x \in D(A^*). \tag{4.4.1}$$

Equation (4.4.1) implies $\ker A^* \subseteq (\mathrm{Ran}A)^\perp$. On the other side, by eq. (4.4.1) $x \in (\mathrm{Ran}A)^\perp$ implies $x \in D(A^*)$ and $A^*x = 0$, i.e. $x \in \ker A^*$; thus, $(\mathrm{Ran}A)^\perp \subseteq \ker A^*$. In conclusion

$$\ker A^* = (\mathrm{Ran}A)^\perp. \tag{4.4.2}$$

B is an *extension* of A, briefly $A \subseteq B$, if $D(A) \subseteq D(B)$ and $A = B$ on $D(A)$; if $D(A)$ is dense $B^* \subseteq A^*$. A densely defined operator A is *hermitian* or *symmetric* if

$$(Ax, y) = (x, Ay) \quad \forall x, y \in D(A); \tag{4.4.3}$$

then $A \subseteq A^*$. An operator A is *closed* if, whenever $D(A) \ni x_n \to x$ and $Ax_n \to y$, then $x \in D(A)$ and $Ax = y$. It follows easily that A^* is closed, since $D(A^*) \ni x_n \to x$ and $A^*x_n \to y$ implies $\forall z \in D(A)$

$$(A\,z, x) = \lim(A\,z, x_n) = \lim(z, A^*x_n) = (z, y), \tag{4.4.4}$$

---

[14]The Heisenberg commutation relations hold on the dense domain of C_0^∞ functions, but there is no common dense domain of analytic vectors for q and p and one has modified Weyl relations in agreement with the physical interpretation. E.g., $e^{i\beta\,p_{\theta=0}}$ describes a translation $x \to x + \beta$ mod 1 and one has $e^{i\beta p_0}\,e^{i\alpha x}\,e^{-i\beta p_0} = e^{i\alpha(x+\beta\ mod\ 1)}$, (see Sect. 6.8 for more details).

i.e. $x \in D(A^*)$ and, by definition of the adjoint, $A^*x = y$. An operator is *closable* if it has a closed extension; the *closure* $\bar{A}$ of A is the smallest closed extension. Thus, any hermitian operator is closable and

$$A \subseteq \bar{A} \subseteq A^*. \tag{4.4.5}$$

A hermitean operator A is *self-adjoint* if $A = A^*$. A is *essentially self-adjoint* if its closure $\bar{A}$ is self-adjoint. The self-adjointness domain D_A contains a dense set D of vectors, called *analytic vectors*, on which the exponential series of e^{tA}, $t > 0$, converges strongly; A is essentially self-adjoint on D (see Reed and Simon book Sect. X.6).

As a preliminary material for the discussion of Kato's theorems, we briefly discuss some criteria of self-adjointness and a characterization of the self-adjoint extensions of a closed symmetric operator.[15]

Theorem 4.4.1 *A hermitian operator A, acting on a Hilbert space $\mathcal{H}$, is self-adjoint iff*

$$\text{Ran}(A \pm i) = \mathcal{H}. \tag{4.4.6}$$

Proof. Let A be self-adjoint. Then, $\ker(A \mp i) = \{0\}$: in fact, $\forall x \in D(A) = D(A^*)$, $(x, Ax) = (Ax, x) = \overline{(x, Ax)}$ is real and $0 = (x, (A \mp i)x)$ implies $(x, x) = 0$. Thus, $\ker(A \mp i) = \{0\}$ and by eq. (4.4.2), applied to the operator $A \pm i$, one gets $\overline{\text{Ran}(A \pm i)} = \mathcal{H}$. To get eq. (4.4.6) it remains to prove that $\text{Ran}(A \pm i)$ is closed: in fact, since $\forall x_n \in D(A)$,

$$||(A \pm i)(x_n - x_m)||^2 = ||A(x_n - x_m)||^2 + ||x_n - x_m||^2, \tag{4.4.7}$$

the strong convergence of $(A \pm i)x_n$ implies that of Ax_n and of x_n and since A is closed,

$$\lim (A \pm i)x_n = (A \pm i) \lim x_n, \quad \text{i.e. } \text{Ran}(A \pm i) \text{ is closed.}$$

Conversely, eq. (4.4.6) implies that $\forall x \in D(A^*)$ there exists $y \in D(A)$ such that

$$(A^* - i)x = (A - i)y = (A^* - i)y, \tag{4.4.8}$$

(since $A \subseteq A^*$); then, since, by eqs. (4.4.2), (4.4.6), $\ker(A^* - i) = \{0\}$, eq. (4.4.8) implies $x = y$, i.e. $D(A^*) = D(A)$ and A is self-adjoint.

Theorem 4.4.2 *A hermitian operator A is essentially self-adjoint iff*

$$\overline{\text{Ran}(A \pm i)} = \mathcal{H}. \tag{4.4.9}$$

Proof. By using eq. (4.4.7) as above one has

$$\text{Ran}(\bar{A} \pm i) = \overline{\text{Ran}(A \pm i)}$$

and the statement follows from the preceding theorem applied to $\bar{A}$.

---

[15] For a general discussion see the above quoted book by Reed and Simon and N. Dunford and J. Schwartz, *Linear Operators*, Interscience 1958, Sect. XII.4.

In general, given a symmetric operator A, the lack of fulfillment of condition (4.4.6) is characterized by

$$n_\pm \equiv \dim \mathrm{Ran}(A \pm i)^\perp = \dim \ker(A^* \mp i), \qquad (4.4.10)$$

which are called the *deficiency indices* of A.

Theorem 4.4.3 *A closed symmetric operator A has self-adjoint extensions iff $n_+ = n_-$. Its self-adjoint extensions B are characterized in the following way*

$$D(B) = \{y; y = y_A + y_+ + U_B y_+, \ y_A \in D(A), \ y_+ \in ker(A^* - i)\},$$

with U_B a unitary operator: $ker(A^* - i) \to ker(A^* + i)$, and

$$By = Ay_A + i\,y_+ - i\,U_B\,y_+.$$

Given a hermitian differential operator A, $(A \subseteq A^*)$, typically defined on the set of C^∞ functions of compact support, a way to get an essentially self-adjoint extension is to extend it to L^2 functions which have sufficient L^2 derivatives and possibly specify boundary conditions so that the differential equation

$$(A^* \pm i)\,\psi = 0$$

has no solution in L^2, (for illustration see the example of the operator $-id/dx$ in $L^2([0,1])$).

We can now prove the Kato-Rellich theorems.

Theorem 4.4.4 *Let H_0 be a self-adjoint operator on $D(H_0)$ and H_1 a symmetric operator smaller than H_0 in the sense of Kato, then $H = H_0 + H_1$ is self-adjoint on $D(H_0)$.*

Proof. It suffices to prove the self-adjointness of H/λ for λ large enough, equivalently, by Theorem 4.4.1, that $\mathrm{Ran}(H \pm i\,\lambda) = \mathcal{H}$. This follows easily from

$$B \equiv H_1(H_0 \pm i\,\lambda)^{-1}$$

being well defined with $||B|| < 1$. In fact, $||B|| < 1$ is a necessary and sufficient condition for the convergence of the Neumann series

$$\sum_{n=0}^{\infty} (-B)^n = (1 + B)^{-1}, \qquad (4.4.11)$$

and the identity

$$H \pm i\,\lambda = (\mathbf{1} + B)\,(H_0 \pm i\,\lambda)$$

implies that $\mathrm{Ran}(H \pm i\,\lambda) = \mathcal{H}$, since $\mathrm{Ran}(H_0 \pm i\,\lambda) = \mathcal{H}$ and $\mathbf{1} + B$ is invertible.

For the proof of $||B|| < 1$ one exploits the Kato smallness condition. In fact, $\forall \chi \in D(H_0)$ the identity

$$||(H_0 \pm i\,\lambda)\,\chi||^2 = ||H_0\,\chi||^2 + \lambda^2\,||\chi||^2, \quad \lambda \in \mathbf{R}^+,$$

implies $||H_0\,\chi|| < ||(H_0 \pm i\,\lambda)\,\chi||$, $||\chi|| < \lambda^{-2}\,||(H_0 \pm i\,\lambda)\,\chi||$, so that, $\forall \psi \in \mathcal{H}$, $\chi = (H_0 \pm i\lambda)^{-1}\,\psi \in D(H_0)$ (since $H_0(H_0 \pm i\lambda)^{-1}$ is bounded) and the Kato smallness condition gives

$$||H_1\,(H_0 \pm i\lambda)^{-1}\,\psi|| \le a\,||H_0\,\chi|| + b\,||\chi||$$

$$\le (a + b\,\lambda^{-1})\,||(H_0 \pm i\,\lambda)\,\chi|| < ||\psi||,$$

since $a < 1$ and for λ large enough $a + b\lambda^{-1} < 1$.

Theorem 4.4.5 *Let $V = V_1 + V_2$, $V_1 \in L^2(\mathbf{R}^3)$, $V_2 \in L^\infty(\mathbf{R}^3)$ be a real potential, then V is smaller than $H_0 = -\Delta/2m$ on $D(H_0)$ (the domain of self-adjointness of H_0) and*

$$H = -\Delta/2m + V$$

is self-adjoint on $D(H_0)$.

Proof. Clearly $D(V) \supset C_0^\infty(\mathbf{R}^3)$ and for any given $\alpha > 0$, $\exists \beta > 0$ such that $\forall \psi \in C_0^\infty(\mathbf{R}^3)$

$$||\psi||_\infty \le \alpha\,||\Delta\,\psi|| + \beta\,||\psi||. \tag{4.4.12}$$

In fact, putting $h(k) \equiv (1 + k^2/2m)^{-1}$, one has

$$(2\pi)^{3/2}||\psi||_\infty \equiv (2\pi)^{3/2}\sup|\psi| = \sup|\int d^3\,k\,e^{i\,kx}\,\tilde{\psi}(k)|$$

$$\le \int d^3k\,|\tilde{\psi}(k)| \equiv ||\tilde{\psi}||_1 \le ||h||\,||\tilde{\psi}\,h^{-1}||$$

$$= C\,||(H_0 + 1)\psi|| \le C\,(||H_0\,\psi|| + ||\psi||).$$

Now, putting $\tilde{\psi}_\lambda(k) \equiv \lambda^3\,\tilde{\psi}(\lambda k)$, $\lambda \in \mathbf{R}^+$, one has $||\tilde{\psi}||_1 = ||\tilde{\psi}_\lambda||_1$, $||\tilde{\psi}|| = \lambda^{-3/2}||\tilde{\psi}_\lambda||$ and

$$(2\pi)^{3/2}||\psi||_\infty \le ||\tilde{\psi}||_1 = ||\tilde{\psi}_\lambda||_1 \le C\,\lambda^{-1/2}\,||H_0\,\psi|| + C\,\lambda^{3/2}||\psi||,$$

i.e. eq. (4.4.12) holds with α as small as one likes. Then, one has

$$||V\psi|| \le ||V_1||\,||\psi||_\infty + ||V_2||_\infty\,||\psi||$$

$$\le 2m\,\alpha||V_1||\,||H_0\,\psi|| + (\beta||V_1|| + ||V_2||_\infty)\,||\psi||,$$

i.e. V is smaller than H_0 on $C_0^\infty(\mathbf{R}^3)$ and clearly it remains so also on the domain of self-adjointness of H_0. [16] The self-adjointness of H follows from Theorem 4.4.4.

We conclude this Appendix with the characterization of self-adjoint operators as generators of one-parameter strongly continuous groups of unitary operators (*Stone Theorem*).

First we remark that the spectral theorem proved in Appendix C extends to unbounded operators. In fact, if F is a finite (but not necessarily bounded) Borel function of a bounded self-adjoint operator A, the r.h.s. of eq. (1.6.10) still defines an operator $F(A)$ with domain

$$D_F = \{x \in \mathcal{H}; \int |F(\lambda)|^2 d(x, P(\lambda)x) < \infty\},$$

since $\forall x \in D_F$ the sequence of vectors

$$F(A)_K \, x \equiv \int_{|F(\lambda)| \leq K} F(\lambda) \, dP(\lambda) x = \int F(\lambda) \chi(F^{-1}([0, K))) \, dP(\lambda) x$$

converges strongly as $K \to \infty$, thanks to the characterization of D_F

$$||(F(A)_{K+J} - F(A)_K) \, x||^2 = \int_{K \leq |F(\lambda)| \leq K+J} |F(\lambda)|^2 d(x, P(\lambda) x) \to 0.$$

Now, given a self-adjoint (unbounded) operator A, $\forall \lambda_0 \in \mathbf{C}$, $\lambda_0 \notin \sigma(A)$, its *resolvent*

$$R(\lambda_0) \equiv (\lambda_0 - A)^{-1}, \tag{4.4.13}$$

is a normal bounded operator, with a spectral representation

$$R(\lambda_0) = \int_{\sigma(R(\lambda_0))} \lambda \, dP(\lambda).$$

Since $F(\lambda) \equiv \lambda_0 - \lambda^{-1}$ for $\lambda \neq 0$ and $F(0) \equiv 0$ is a finite Borel function, by the above argument the corresponding spectral integral defines an operator which coincides with $A = \lambda_0 - R(\lambda_0)^{-1}$ on D_F. Finally, in terms of the new measure

$$\mu(B) \equiv P(F^{-1}(B)), \quad \forall B \in \mathcal{B}(\mathbf{R})$$

---

[16] Since $C_0^\infty \subset D(H_0)$ any $\psi \in D(H_0)$ defines a continuous linear functional on C_0^∞, i.e. a Schwartz distribution, and as such satisfies

$$(\varphi, H_0 \, \psi) = (-\Delta \varphi, \psi) = (\varphi, -\Delta \psi), \quad \forall \varphi \in C_0^\infty.$$

Thus, the domain of the adjoint of $-\Delta$ consists of all the $\psi \in L^2$, such that the distributional derivative $\Delta \psi$ belongs to L^2. Clearly, this is also the domain of self-adjointness of H_0.

one gets the standard spectral representation of A on a suitable (dense) domain

$$A = \int_{\sigma(A)} \lambda \, d\mu(\lambda). \tag{4.4.14}$$

The above spectral representation allows to extend Theorem 1.6.8 to the case of an unbounded self-adjoint operator A, with the derivative taken on vectors of the self-adjointness domain of A.

Conversely, given a one-parameter group of strongly continuous unitary operators $U(t)$, the generator A is defined by

$$A\psi \equiv i \, d[U(t)\psi]/dt,$$

on the domain

$$D_A \equiv \{\psi \in \mathcal{H}; U(t)\psi \text{ is differentiable in } t, \text{at } t = 0\}. \tag{4.4.15}$$

Clearly, by the unitarity of $U(t)$, $\forall \psi, \phi \in D_A$, $(\psi, A\phi) = (A\psi, \phi)$, i.e. A is symmetric.

Furthermore, A is densely defined, since i) D_A contains the $U(t)$ invariant (Gårding) domain G, generated by the vectors of the form $\varphi_f = \int dt \, f(t) \, U(t) \varphi$, $f \in \mathcal{D}(\mathbf{R})$, which are well defined thanks to the strong continuity of $U(t)$, and on which $A\varphi_f = -i \, \varphi_{f'}$, and ii) G is dense because $\forall \varphi \in \mathcal{H}$, $\varphi_{f_n} \to \varphi$, whenever f_n is a smooth approximation of the Dirac δ, as $n \to \infty$.

Moreover, $\psi_\pm \in \mathrm{Ran}(A \pm i)^\perp$ implies, $\forall \chi \in G$,

$$\frac{d}{dt}(\psi_\pm, U(t)\chi) = \pm(\psi_\pm, U(t)\chi) \equiv \pm F_\pm(t)$$

i.e. $F(t) = F(0) \, e^{\pm t}$, which is incompatible with the boundedness of $U(t)$, unless $F(0) = 0$. Hence, by Theorem 4.4.2 A is essentially self adjoint on G.

Finally, if $\bar{A}$ denotes the unique self-adjoint extension of A and $V(t) \equiv e^{i\bar{A}t}$, $\forall \psi, \phi \in G$ one has

$$\frac{d}{dt}(\psi, V(t)^* U(t) \phi) = \frac{d}{dt}(V(t)\psi, U(t)\phi) = 0,$$

i.e. $V(t)^* U(t) = \mathrm{const} = 1$ and $A = \bar{A}$.

Chapter 5

Examples

5.1 Double-slit interference and particle-wave duality

To clarify the particle-wave duality briefly mentioned in Sect. 1.1, we consider the following idealized experiment (a realistic one based on the same mechanism would require a more sophisticated experimental setting).

We consider an impenetrable plane Π with two circular slits or holes of radius δ and at a distance d from it a parallel plane acting as a screen S. Π can be taken as the $x - y$ plane, with holes centered along the x-axis at $x_1 = -a$ and $x_2 = a$, respectively and S as the plane $z = d$. Electrons, all with (approximately) the same energy, are shot from below the plane Π and, by means of a detector on the screen, one measures the probability distribution of electrons arriving at the screen. For classical particles one would get a probability distribution $P(x, y, d)$ which is the sum of the probabilities P_1, P_2 corresponding to the cases in which only one of the two holes is open:

$$P(x, y, d) = P_1(x, y, d) + P_2(x, y, d). \tag{5.1.1}$$

For a quantum particle, this is no longer the case and one gets an extra interference term $P_{12}(x, y, d)$ which, depending on the point (x, y) on the screen, may act constructively or destructively, with the result that P will be greater or smaller than $P_1 + P_2$. This phenomenon is typical of a wave-like behaviour and, in fact, interference patterns are observed if light is sent on the plane (Young experiment).

A better understanding of the wave behaviour of a quantum particle can be obtained by actually computing the probability distribution P according to the laws of Quantum Mechanics. If at the initial time $t = 0$, the electron wave function is the sum of two Gaussians centered at the two holes

$$\psi(\mathbf{x}, 0) = \psi_1(\mathbf{x}, 0) + \psi_2(\mathbf{x}, 0),$$

$$\psi_i(\mathbf{x}, 0) = C \exp\left[-(x - x_i)^2/(2\Delta_0 x)^2\right] e^{-y^2/(2\Delta_0 y)^2} e^{i<p_z>z} e^{-z^2/(2\Delta_0 z)^2}$$

$$\equiv \phi(y, z)\, \phi_i(x), \quad i = 1, 2,$$

with C a suitable normalization constant, $\Delta_0 x, \Delta_0 y, \Delta_0 z$ smaller than δ, $< p_z >> 0$, then at a later time t the wave function is the superposition of two wave functions given by formulas similar to those of Sect.4.2. The probability distribution on the screen $z = d$, for $t \approx m\,d/ < p_z >$, (m being the electron mass) is then given by

$$P(x, y, d, t) = |\psi(x, y, d, t)|^2$$

$$= |\phi(y, d, t)|^2 \left(|\phi_1(x, t)|^2 + |\phi_2(x, t)|^2 + 2\,\mathrm{Re}(\bar{\phi}_1(x, t)\, \phi_2(x, t))\right)$$

and one observes the occurrence of interference terms, contrary to classical probability theory.[1] It should be noted that such an interference phenomenon occurs even if only one electron is sent at a time towards the double slit plane, (self-interference).

5.2 The quantum harmonic oscillator. Energy quantization

In analogy with the classical case, the (one dimensional) quantum harmonic oscillator is defined by the following Hamiltonian

$$H = \frac{p^2}{2m} + \frac{m\omega^2}{2}\, q^2. \tag{5.2.1}$$

In the classical case, the possible values of the energy form a continuum set; we want to show that in the quantum case only discrete values are possible. For this purpose, according to the general discussion about the relation between spectrum and expectation values (see Sect. 1.3, Appendix C, Prop. 1.6.4) we have to determine the spectrum of the Hamiltonian, which describes the quantum energy.

According to the discussion of Sects. 3.3, 4.3, one can work out the problem in the Schroedinger representation; since the potential satisfies the condition of Theorem 4.3.4,

$$H = -\hbar^2 \Delta/2m + m\omega^2 x^2/2$$

---

[1] For a nice discussion of the above (idealized) experiment and its deep philosophical implications see R.P. Feynman, R.B. Leighton and M. Sands, *The Feynman Lectures on Physics*, Vol. I, Addison-Wesley 1963, Chaps. 37, 38.

is essentially self-adjoint on $C_0^\infty(\mathbf{R})$ and its spectrum can be determined by Hilbert space techniques.

We will rather determine the spectrum by algebraic methods. For this purpose, to simplify the computations, we perform a change of variables $q' = (m\,\omega/\hbar)^{1/2}\,q$, $p' = (m\hbar\omega)^{-1/2}\,p$, $H' = (\hbar\omega)^{-1}\,H$ and introduce the so-called destruction and creation (unbounded) operators

$$a = (q' + i\,p')/\sqrt{2}, \quad a^* = (q' - ip')/\sqrt{2}, \tag{5.2.2}$$

which obey the following commutation relations

$$[a, a^*] = 1. \tag{5.2.3}$$

Then, the Hamiltonian H' can be written as

$$H' = (1/2)(a^*\,a + a\,a^*) = a^*\,a + 1/2 \equiv N + 1/2$$

and one has

$$[N, a] = -a, \quad [N, a^*] = a^*.$$

The point is to prove that the spectrum of N is discrete. In a certain sense, the following analysis can be viewed as an alternative proof of Von Neumann theorem, once one has disposed of the technical point that in a regular representation the operators q and p have a common invariant dense domain D of analytic vectors for $p^2 + q^2$. [2]

Since N is self-adjoint, one may define the unitary operator $T(s) \equiv \exp{(isN)}, s \in \mathbf{R}$ and prove that (on D)

$$T(s)\,a\,T(s)^{-1} = e^{-is}\,a, \quad T(s)\,a^*\,T(s)^{-1} = e^{is}\,a^*. \tag{5.2.4}$$

In fact, $F(s) \equiv T(s)\,a\,T(s)^*$ satisfies

$$dF(s)/ds = i\,e^{isN}\,[N, a]\,e^{-isN} = -iF(s),$$

which implies $F(s) = \exp{(-is)}\,F(0)$, so that the above equation holds.

Equations (5.2.4) imply that $T(2\pi)$ commutes with a, a^* and therefore with the (Heisenberg) algebra generated by them; by irreducibility, $T(2\pi)$ must be a multiple of the identity, say $T(2\pi) = \exp{i\theta\mathbf{1}}$, and therefore $T'(s) \equiv T(s)\exp{(-is\theta/2\pi)}$ satisfies $T'(2\pi) = T'(0) = \mathbf{1}$. Then, the spectral representation of $T'(s)$

$$T'(s) = \int_{\sigma(N')} dE(\lambda)\,e^{is\lambda}, \quad N' \equiv N - \theta/2\pi$$

---

[2] This follows quite generally, as proved in O. Bratteli and D.K. Robinson, *Operator Algebras and Quantum Statistical Mechanics*, Vol. II, Springer 1981, Sect. 5.2.3; it is easy to check it directly in the Schroedinger representation.

gives

$$0 = (T'(2\pi) - \mathbf{1})^2 = \int_{\sigma(N')} dE(\lambda)\,(e^{i2\pi\,\lambda} - 1)^2,$$

so that supp $dE(\lambda) \subseteq \mathbf{Z}$, i.e. the spectrum of N' and therefore of N and of H' is discrete (*energy quantization*).

One can compute the spectrum explicitly. Let $0 < \lambda \in \sigma(N)$ and Ψ_λ a corresponding eigenvector; then $a\Psi_\lambda$ is well defined and $\neq 0$, since

$$||a\,\Psi_\lambda||^2 = (\Psi_\lambda, N\,\Psi_\lambda) = \lambda||\Psi_\lambda||^2 > 0,$$

and one has

$$T(s)a\Psi_\lambda = T(s)aT(-s)e^{is\lambda}\,\Psi_\lambda = e^{i(\lambda-1)s}a\Psi_\lambda.$$

The above equation says that also $\lambda - 1 \in \sigma(N)$. Since the argument applies to any $\lambda > 0$, this is compatible with the positivity of $\sigma(N)$ only if $0 \in \sigma(N)$ and $a\Psi_0 = 0$. Actually $\sigma(N) = \{n, \ n \in \mathbf{N}\}$ since a similar argument with a replaced by a^* shows that $a^*\,\Psi_\lambda$ has eigenvalue $\lambda + 1$. This explains the name of destruction and creation operators since they decrease, respectively increase, the energy. In conclusion, the spectrum of the Hamiltonian is

$$\sigma(H) = \{E_n = \hbar\,\omega\,(n + 1/2), \quad n \in \mathbf{N}\}. \tag{5.2.5}$$

The eigenvector Ψ_0 corresponding to $n = 0$ is the Fock vector introduced in Sect. 3.2; this follows easily from the fact that $N\Psi_0 = 0$ implies $||a\Psi_0||^2 = (\Psi_0, N\Psi_0) = 0$, i.e. $a\Psi_0 = 0$, which coincides with eq. (3.4.3). In the Schroedinger representation, the Fock condition amounts to the following differential equation

$$\sqrt{2}a\Psi_0(x') = (x' + d/dx')\Psi_0(x') = 0,$$

whose solution is the Gaussian wave function discussed in Sect. 3.4. The eigenfunctions corresponding to $n \neq 0$ are obtained by applying a^* n-times

$$\Psi_n(x') = (n!)^{-1/2}\,(a^*)^n\Psi_0(x') = (n!\,2^n)^{-1/2}e^{-x'^2/2}\,H_n(x'),$$

where $H_n(x')$ are the Hermite polynomials of degree n and parity $(-1)^n$.

It is interesting to note that for a classical harmonic oscillator of given energy E, only the configurations for which $V(x) = m\omega^2x^2/2 \leq E$ are accessible, whereas for the quantum harmonic oscillator the probability of finding the particle at a point x for which $V(x) > E$ is non-zero. For example, for the harmonic oscillator in the ground state, one has $|\Psi_0(x)|^2 \neq 0$ also for $|x| > \sqrt{2E/m\omega^2}$. This does not mean that there is a violation of the energy conservation since a measurement of the position is inevitably accompanied by a momentum transfer and therefore by a change of the energy.

5.3 Quantum particle in a square potential well and in a box

We consider a quantum particle of mass $m = 1/2$ in one dimension subject to a square potential well (in units with $\hbar = 1$)

$$V(x) = 0, \quad |x| > L; \qquad V(x) = V_0 < 0, \quad |x| \le L,$$

and study the spectrum of the Hamiltonian [3] $H = -(d/dx)^2 + V(x)$.

Since $-(d/dx)^2$ is a positive operator, $H \ge V(x) \ge \operatorname{Inf} V(x) = V_0$ and therefore $E \in \sigma(H)$ only if $E \ge V_0$. Furthermore, $E > 0$ belongs to the continuous spectrum, since in this case the eigenvalue equation

$$-(d/dx)^2 \, \psi(x) = (E - V(x)) \, \psi(x), \quad \psi \in D(-\Delta),$$

implies that, for $|x| > L$, $\psi(x) \sim A_1 \exp(i\sqrt{E}x) + A_2 \exp(-i\sqrt{E}x) \notin L^2$. Such a continuous spectrum is doubly degenerate.

We now discuss the region $V_0 \le E \le 0$. Since the potential is even, if $\psi(x)$ is a solution, so is $\psi(-x)$ and one may consider solutions with definite parity: $\psi_\pm(x) = \pm \psi_\pm(-x)$. One has

$$\psi_\pm(x) = A \, e^{\rho x}, \quad \rho \equiv |E|^{1/2}, \quad x < -L,$$

$$\psi_\pm(x) = B(e^{ikx} \pm e^{-ikx}), \quad k \equiv \sqrt{E - V_0}, \quad |x| < L.$$

The condition that $\psi \in D(-\Delta)$ requires that $\psi_\pm$ be continuous together with their derivatives at the points $x = -L$, $x = L$. Thus, one gets for the parity even solutions ψ_+,

$$A/2B = e^{\rho L} \cos(kL), \quad \rho A/2B = e^{\rho L} \, k \sin(kL),$$

which imply $k \operatorname{tg}(kL) = \rho$. Similarly, for the parity odd solutions one gets $k \operatorname{cotg}(kL) = -\rho$. These equations may be easily solved by graphical methods and one finds that there are discrete solutions (at least one if $V_0 < E < 0$ and no solution if $E = V_0$; $E = 0$ would lead to non-L^2 behaviour). Thus, the energy spectrum for $V_0 < E < 0$ is discrete (*energy quantization*). The corresponding eigenfunctions are essentially localized inside the potential well and clearly remain so at any time (*bound states*).

By a similar method one can determine the energy spectrum when

$$V(x) = V_L > 0, \quad |x| > L, \qquad V(x) = 0, \quad |x| \le L.$$

---

[3] For the study of the spectrum of the Schroedinger operators see the references given in the Introduction and e.g. A. Galindo and P. Pascual, *Quantum Mechanics I*, Springer 1990.

Now the points of the spectrum must satisfy $E \geq V(x) \geq \mathrm{Inf}\, V(x) = 0$ and the spectrum is continuous for $E > V_L$. For $E < V_L$ one gets the same equations as before, with however $k \equiv \sqrt{E}$, $\rho \equiv |E - V_L|^{1/2}$.

The eigenvalue spectrum becomes very simple if we let $V_L \to \infty$, i.e. if we consider a quantum particle inside an impenetrable box of size $2L$. In this case, the equation for the parity even solution requires $\mathrm{tg}(LE^{1/2}) = \infty$, i.e. $\sqrt{E} = (2m + 1)\pi/2L$, $m \in \mathbf{N}$, whereas for the parity odd solutions one has $\mathrm{tg}(LE^{1/2}) = 0$, i.e. $\sqrt{E} = 2m\pi/2L$. In conclusion the spectrum consists of the points

$$E_n = (\pi/2L)^2 \, n^2, \quad n \in \mathbf{N}. \tag{5.3.1}$$

Clearly, there is no continuous spectrum (*quantization of the energy*).

In all the cases discussed above, the quantum behaviour significantly differs from the classical one.[4] In particular, for a bounded potential there is a non vanishing probability of finding the quantum particle in regions which are energetically forbidden in the classical case, as discussed for the quantum oscillator (*tunnel effect*).

In the case $V_L \to \infty$, the wave function is non-zero only inside the box $-L < x < L$ and it vanishes at the boundaries $|x| = L$ (but its derivative does not). The point is that the limiting potential $V_L = \infty$ does not satisfy Kato criterium so that the differential operator $H = -(d/dx)^2 + V(x)$ does not define a (unique) self-adjoint operator on $D(-(d/dx)^2)$ (see Appendix F). In fact, in this limit the model describes a free quantum particle confined to live in the interval $-L < x < L$, and the differential operator $-(d/dx)^2$, which is well defined on $D_0 \equiv C_0^\infty(-L, L)$, does not define a time evolution unless one specifies one of its self-adjoint extensions, which are parametrized by the boundary conditions for ψ and its first derivatives.

---

[4]For a nice discussion of potential well problems, also as models of physical systems, see J.-M. Lévy-Leblond and F. Balibar, *Quantics*, North-Holland 1990, Chap. 6.

5.4 Quantization of the angular momentum. The spin

In the previous sections we discussed the quantum version of the position, momentum and energy and found significant differences with respect to the classical case; another basic quantity is the angular momentum and we shall discuss its quantum properties in this section. We consider a quantum particle and its (orbital) angular momentum defined in analogy with the classical case as

$$L_j = \varepsilon_{jkl} x_k p_l, \qquad (5.4.1)$$

where $j, k, l = 1, 2, 3$, ε_{jkl} is the totally antisymmetric tensor, $\varepsilon_{123} = 1$, and sum over repeated indices is understood, as usual. In constructing the quantum version of classical physical quantities as functions of q and p an ambiguity arises because the order of the variables is not irrelevant in the non abelian case; here, however, there is no problem because by the Heisenberg commutation relations $\varepsilon_{jkl} x_k p_l = \varepsilon_{jkl} p_l x_k$.

By using the commutation relations between x_k and p_l and the identity $[A, BC] = [A, B]C + B[A, C]$, we get

$$[L_j, L_k] = i\hbar \, \varepsilon_{jkl} L_l. \qquad (5.4.2)$$

For simplicity, in the following discussion, we will put $\hbar = 1$ by a suitable choice of units. One recognizes in the above equations the Lie algebra relations of the generators of $SO(3)$. Since $SO(3)$ is a compact Lie group, its irreducible unitary representations in a Hilbert space are finite dimensional and the spectrum of the generators is discrete.[5] This means that the quantum (orbital) momentum is *quantized*. Due to the physical relevance of this result, taken for granted the technical fact that for unitary representations there is a common dense domain of analytic vectors for all the generators,[6] we present an explicit derivation, which also gives the form of the spectrum in each irreducibile representation.

Putting $L^2 \equiv \sum_i L_i^2$, one easily sees that the commutation relations (5.4.2) imply

$$[L_i, L^2] = \varepsilon_{ijk}(L_k L_j + L_j L_k) = 0, \qquad (5.4.3)$$

so that, by irreducibility, L^2 must be a multiple of the identity, say $\lambda^2 \mathbf{1}$, since L^2 is a positive operator.

The case $\lambda = 0$ requires $L_i^2 = 0$, $\forall i$, i.e. $L_i = 0$, i.e. the trivial representation. We then take $\lambda > 0$ and determine the spectrum of anyone

---

[5]Quite generally, representations of compact topological groups can be reduced to unitary representations and the irreducible ones are finite dimensional; see e.g. A.O. Barut and Rączka, *Theory of Group Representations and Applications*, World Scientific 1986, Chap. 6.

[6]E. Nelson, Ann. Math. **70**, 572 (1959); see also the above quoted book by Barut and Rączka, Chap. 11, Sect. 4.

of the generators, say L_3. Since $L^2 \geq L_3^2$, we have $\sigma(L_3) \subseteq [-\lambda, \lambda]$, i.e. the spectrum is bounded from below and from above. Thus, all the L_i's are bounded. Furthermore, putting $L_\pm \equiv L_1 \pm i L_2$, one has

$$[L_3, L_\pm] = \pm L_\pm, \quad e^{is L_3} L_\pm e^{-is L_3} = e^{\pm is} L_\pm. \tag{5.4.4}$$

The last equation implies that $e^{i2\pi L_3}$ commutes with all the generators and therefore by irreducibility is a multiple of the identity; by an argument similar to that used in Sect. 5.2, one concludes that the spectrum of L_3 is discrete (*angular momentum quantization*).

For the explicit determination of the spectrum, let $l_3 \in \sigma(L_3)$ and Ψ_{l_3} the corresponding eigenvector, then

$$L_3 L_\pm \Psi_{l_3} = (l_3 \pm 1) L_\pm \Psi_{l_3},$$

so that, if $L_\pm \Psi_{l_3} \neq 0$, also $l_3 \pm 1$ is an eigenvalue and $L_\pm$ have the meaning of raising and lowering operators. Since the spectrum of L_3 is bounded, the sequence $l_3 \pm n$ must terminate, i.e. there must be a maximal value $l_3 = l$ such that $L_+ \Psi_l = 0$ and a minimal value $l_3 = k$ such that $L_- \Psi_k = 0$; furthermore $l - k \in \mathbf{N}$. Since

$$0 = (\Psi_l, L_- L_+ \Psi_l) = (\Psi_l, (L^2 - L_3^2 - L_3)\Psi_l) = \lambda^2 - l(l+1),$$

$$0 = (\Psi_k, L_+ L_- \Psi_k) = (\Psi_k, (L^2 - L_3^2 + L_3)\Psi_k) = \lambda^2 - k(k-1),$$

one has $\lambda^2 = l(l+1) = k(k-1)$. This equation implies $l = \lambda^2 - l^2 \geq 0$, $-k = \lambda^2 - k^2 \geq 0$ and has only two solutions: $k = -l$ and $k = l+1$ and only the first is acceptable, because, by definition, $k < l$. Thus, $l - k = 2l$ is an integer and the spectrum consists of the $2l+1$ points $-l, -l+1, ..., l-1, l$.

The above derivation of the spectrum of the angular momentum has been obtained by using only the commutation relations (5.4.2); now we fully exploit the consequences of eq. (5.4.1). By using polar coordinates for the arguments of the wave functions, $x_1 = r \sin\theta \cos\phi$, $x_2 = r \sin\theta \sin\phi$, $x_3 = r \cos\theta$, we have the following representation

$$L_3 = -i\, \partial/\partial\phi, \quad L_\pm = e^{\pm i\phi}(i \cot\theta\, \partial/\partial\phi \pm \partial/\partial\theta)$$

and the condition that $\psi(x_1, x_2, x_3) = \hat{\psi}(r, \theta, \phi)$ is an eigenfunction of L_3 implies that

$$\hat{\psi}(r, \theta, \phi) = e^{il_3\, \phi} \hat{\psi}(r, \theta).$$

The requirement that $\hat{\psi}$ be single valued, $\hat{\psi}(r, \theta, \phi + 2\pi) = \hat{\psi}(r, \theta, \phi)$, further implies that l_3 is an integer, say m, and so is l. The explicit form of $\psi(r, \theta)$ is obtained by solving the eigenvalue equation for L^2. The solutions are given by the so-called spherical harmonics $Y_{lm}(\theta, \phi)$:

$$\hat{\psi}(r, \theta, \phi) = \psi(r)\, Y_{lm}(\theta, \phi), \tag{5.4.5}$$

with $\psi(r)$ an undetermined function. Clearly, if $L^2\psi = l(l+1)\,\psi$, $L_3\,\psi = l\psi$, then $L^2\,e^{i\phi}\psi = (l+1)(l+2)\,e^{i\phi}\psi$. Thus the spectrum of L^2 in $L^2(\mathbf{R}^3)$ contains all the points $l(l+1)$, $l \in \mathbf{N}$.

It is not difficult to see that $U_3(\alpha) \equiv e^{-i\alpha L_3}$ implements the rotation of angle α around the third axis, in the sense that

$$(U_3(\alpha)\psi)(r,\theta,\phi) = \psi(r,\theta,\phi-\alpha).$$

Equivalently, $U_3(\alpha)$ induces a rotation of angle α on the operators, through the equation $A' = U_3(\alpha)\,A\,U_3(\alpha)^*$; it is easy to check these transformations, e.g. for small α, for the operators $\mathbf{x}$ or $\mathbf{p}$. Quite generally,

$$U(\mathbf{n},\alpha) \equiv e^{-i\alpha\mathbf{L}\cdot\mathbf{n}}$$

implements the rotations of angle α around the direction $\mathbf{n}$, so that if A is a scalar function, then $[L_j, A] = 0$. On the other hand, if $\mathbf{A}$ is a vector function of $\mathbf{x}$ and $\mathbf{p}$, its transformation properties yield

$$[L_j, A_k] = i\,\varepsilon_{jkl}A_l. \tag{5.4.6}$$

Equations (5.4.6) imply that if $L^2\psi = l(l+1)\,\psi$, $L_3\,\psi = l\psi$, then one has $L^2\,(A_1 + iA_2)\psi = (l+1)(l+2)\,(A_1 + A_2)\psi$. The commutation relations between an operator A and the angular momentum characterize its transformation properties under rotations.

One may wonder whether the representations of the Lie algebra (5.4.2) corresponding to half-integer l have any physical interest. As we have just seen, they cannot describe the orbital angular momentum (5.4.1), but they could describe other forms of angular momentum than that given by (5.4.1). As a matter of fact, there is experimental evidence that most elementary particles have an additional *intrinsic angular momentum* $\mathbf{S}$ called *spin*, which is not a function of q and p, and it rather corresponds to other non strictly mechanical degrees of freedom, like the angular momentum of a spinning top in the pointlike limit. Therefore the spin describes other degrees of freedom and it commutes with q and p. The algebraic relations (5.4.2), which characterize an angular momentum, still hold

$$[S_j, S_k] = i\varepsilon_{jkl}\,S_l, \tag{5.4.7}$$

but the absence of the constraint (5.4.1) allows for half-integer representations, namely those for which the eigenvalue of $\mathbf{S}^2$ is $s(s+1)$ with s half-integer. From a group-theoretical point of view, they correspond to *spinor* representations [7] of the group $SU(2)$, the universal covering of $SO(3)$ (both groups have the same Lie algebra).

A particle with spin is defined by an extended Weyl algebra, including the spin operators, e.g. $S(\gamma) \equiv \exp{(i\gamma \cdot \mathbf{S})}$, $\gamma \in \mathbf{R}^3$. The Hilbert space

---

[7]See e.g. E. Cartan, *The theory of spinors*, Hermann 1960.

of the extended Schroedinger representation is $\mathcal{H} = L^2(\mathbf{R}^d) \times \mathbf{C}^{2s+1}$ and the wave function is a $2s + 1$-component L^2 function, or *spinor*. As a basis in spin space one can take the $2s + 1$ eigenvectors of S_3, $\psi_{s_3}(x)$, $s_3 = -s, -s+1, ..., s$, with $(S_3\psi_{s_3})(x) = s_3\,\psi_{s_3}(x)$; $(S_\pm\psi_{s_3})(x) = \psi_{s_3\pm1}(x)$, for $\pm s_3 < s$, respectively, and $= 0$ otherwise. For a particle with spin, the generator of rotations around $\mathbf{n}$, including the spin degrees of freedom, is the total angular momentum $\mathbf{J} \equiv \mathbf{L} + \mathbf{S}$.

It is useful to work out explicitly the case $s = 1/2$, which covers most of the elementary particles with spin, such as the electron, the proton, the neutron etc.; the spin representation is two-dimensional and it is easy to see that $S_i = (1/2)\sigma_i$, with σ_i the Pauli matrices

$$\sigma_1 = \begin{pmatrix} 0 & 1 \\ 1 & 0 \end{pmatrix}, \qquad \sigma_2 = \begin{pmatrix} 0 & -i \\ i & 0 \end{pmatrix}, \qquad \sigma_3 = \begin{pmatrix} 1 & 0 \\ 0 & -1 \end{pmatrix}.$$

5.5   The Hydrogen atom

As discussed in Sect. 1.1, the basic motivation for the birth of Quantum Mechanics was the description of atoms and the explanation of atomic spectra. The latter ones indicate that, contrary to the classical case, only discrete orbits are allowed, equivalently only discrete values of the energy are permitted; furthermore the stability of the atoms strongly suggests that there is a lowest energy orbit.

We shall see how the theory discussed in the previous sections provide an explanation of these phenomena. The problem amounts to determine the spectrum of the Hamiltonian, which for the hydrogen atom (i.e. an electron bound to a proton) reads

$$H = \frac{p^2}{2m} - \frac{e^2}{r}, \tag{5.5.1}$$

where m is the reduced mass, $m = m_e m_p/(m_e + m_p)$, m_e is the electron mass and $m_p \approx 1800 m_e$ is the proton mass. More generally, for an electron bound to a nucleus of charge Ze, the Coulomb potential is $-Ze^2/r$. To simplify the formulas, we shall use the so-called atomic units, according to which the unit of length is a_0/Z, with $a_0 = \hbar^2/me^2$ (the radius of the first Bohr orbit), the angular momentum is measured in units of $\hbar$, the momentum $\mathbf{p}$ in units $\hbar Z/a_0$ and the energy in units $Z^2 e^2/2a_0$. Then, the Hamiltonian reads

$$H = p^2 - 2/r. \tag{5.5.2}$$

To determine its spectrum one can proceeds by analytic methods, by using the Schroedinger representation; this is an example of analysis of the spec-

trum of Schoedinger operators. [8] We shall rather use algebraic methods.[9]
By an argument similar to that used for the one-dimensional problems, one
can show that $E > 0$ belongs to the continuous spectrum and it does not
describe bound states (similarly to the classical case). We are therefore
interested in the negative spectrum.

First, we remark that the spin does not enter in the potential and will
be neglected. Since H is a scalar function of q and p, it commutes with the
orbital angular momentum

$$[H, \mathbf{L}] = 0. \tag{5.5.3}$$

It is convenient to introduce the quantum version of the Laplace-Runge-
Lenz vector

$$A_i = \frac{x_i}{r} - (1/2)\,\varepsilon_{ijk}(p_j\,L_k + L_k\,p_j) = \frac{x_i}{r} - \varepsilon_{ijk}\,p_j\,L_k + i\,p_i, \tag{5.5.4}$$

which plays an important role in the classical case, where it describes a
vector pointing in the direction of the major axis of the orbit, with a size
equal to the eccentricity, apart from numerical factors; its constancy in time
corresponds to the orbit being closed, a characteristic property of the r^{-2}
central forces. [10]

Since $\mathbf{A}$ is a vector function of $\mathbf{x}, \mathbf{p}$, we have

$$[L_j, A_k] = i\,\varepsilon_{jkl} A_l. \tag{5.5.5}$$

Furthermore, one has

$$\mathbf{L} \cdot \mathbf{A} = \mathbf{A} \cdot \mathbf{L} = 0, \tag{5.5.6}$$

where the first equality follows from the preceding equation and the second
from

$$\mathbf{p} \cdot \mathbf{L} = \mathbf{L} \cdot \mathbf{p} = \varepsilon_{ijk} x_j p_k p_i = 0,$$

$$\mathbf{L} \cdot \mathbf{x} = \mathbf{x} \cdot \mathbf{L} = \varepsilon_{ijk} x_i x_j p_k = 0,$$

$$\mathbf{A} \cdot \mathbf{L} = -\varepsilon_{ijk} p_j L_k L_i = -(1/2)\varepsilon_{ijk} p_j [L_k, L_i] = -i\,\delta_{jl}\,p_j L_l = 0.$$

As in the classical case, $\mathbf{A}$ is a constant of motion, i.e.

$$[H, \mathbf{A}] = 0. \tag{5.5.7}$$

In fact,

$$[H, \mathbf{x}/r] = (-i/r^3)(\mathbf{L} \wedge \mathbf{x} - \mathbf{x} \wedge \mathbf{L})$$

[8]See e.g. P.A.M. Dirac, *The Principles of Quantum Mechanics*, 4th ed., Oxford
Claredon Press, 1958, Sects. 38, 39; A. Galindo and P. Pascual, *Quantum Mechanics*,
Springer 1990.

[9]W. Pauli, Z. Phys. **36**, 336 (1926); English transl. in *Sources of Quantum Mechanics*, B.L. Van der Waerden ed., North-Holland 1967; V. Bargman, Z. Phys. **99**, 576
(1936).

[10]See e.g. W.H. Heintz, Am. Jour. Phys. **42**, 1078 (1974).

and since $2\mathbf{x}/r^3 = -i[H, \mathbf{p}]$ and r^3 and H commute with $\mathbf{L}$, the right hand side becomes

$$-(1/2)(\mathbf{L} \wedge [H, \mathbf{p}] - [H, \mathbf{p}] \wedge \mathbf{L}) = (1/2)[H, (\mathbf{p} \wedge \mathbf{L} - \mathbf{L} \wedge \mathbf{p})].$$

By simple algebraic computations one proves (see Appendix G) that [11]

$$\mathbf{A}^2 = 1 + H + H L^2, \tag{5.5.8}$$

$$[A_i, A_j] = -i\,\varepsilon_{ijk}\, L_k\, H. \tag{5.5.9}$$

Now, $\mathcal{H}_\varepsilon \equiv P_\varepsilon \mathcal{H}$, with P_ε the spectral projection corresponding to the part of the energy spectrum contained in $[-\infty, -\varepsilon]$, $\varepsilon > 0$, is left invariant by $\mathbf{L}$ and $\mathbf{A}$ and on it $(-H)^{-1/2}$ is a well defined operator. We can then introduce the following operators

$$J_i^\pm \equiv (1/2)(L_i \pm (-H)^{-1/2} A_i) \tag{5.5.10}$$

and recognize that as a consequence of above equations we have

$$[J_j^\pm, J_k^\pm] = i\varepsilon_{jkl}\, J_l^\pm, \qquad [J_j^+, J_k^-] = 0, \tag{5.5.11}$$

$$(\mathbf{J}^+)^2 = (\mathbf{J}^-)^2 = -(1/4)(1 + H^{-1}). \tag{5.5.12}$$

Equations (5.5.11) and the first of (5.5.12) are the relations which characterize the Lie algebra of $SO(4)$ and from the above equation we get

$$H = -(1 + 4(\mathbf{J}^+)^2)^{-1}. \tag{5.5.13}$$

Since $J^\pm$ satisfy the commutation relations of two independent angular momenta, by the results of Sect. 5.4, the spectrum of $(J^+)^2$ is discrete and the corresponding eigenvalues are of the form $j(j+1)$, with $2j$ a non negative integer. This implies that the negative spectrum of the Hamiltonian is discrete and the energy eigenvalues have the following form (in atomic units)

$$E = -\frac{1}{(2j+1)^2} = -\frac{1}{n^2} \equiv E_n, \tag{5.5.14}$$

where $n = 2j + 1$ is a positive integer and each eigenvalue is $(2j + 1)^2$ degenerate. In ordinary units one would have $E_n = -m\, Z^2\, e^4/2n^2\hbar^2$. By a more detailed analysis one can show that all positive n occur in the bound state spectrum of the hydrogen atom. [12]

---

[11]The first relation is the quantum version of the classical orbit relation given by Kepler's law of conical sections.

[12]In fact, the eigenspace corresponding to the eigenvalue $n = 2j + 1$ contains states with $J_3^+ = j = J_3^-$ which are therefore eigenstates of $L_3 = J_3^+ + J_3^-$ with eigenvalue $l = 2j$; correspondingly L^2 must have eigenvalue $l'(l' + 1)$, $l' \geq 2j$. Conversely, the eigenspace corresponding to $L^2 = l(l + 1)$, contains vectors with $L_3 = l$; then $J_3^\pm$ must have eigenvalues $j \geq l/2$ and $(\mathbf{J}^\pm)^2$ must have eigenvalues greater than $j(j+1)$, $j \geq l/2$.

The important conclusions of the above result are

i) contrary to the classical case, *the energy spectrum of the bound states* is *discrete*, as if only a discrete set of orbits were allowed

ii) the energy of the bound states is bounded below, i.e. there is a lowest energy state or ground state, corresponding to $n = 1$; this implies stability since an atom in the ground state cannot make transition to a lower energy state when the electromagnetic radiation is taken into account

iii) in the emission or absorption of electromagnetic radiation, associated to transitions from one energy level to another, only discrete amounts of energy are released or absorbed, $(\Delta E)_{nm} = E_n - E_m$, and, by the Einstein relation between energy and frequency of the quanta of the (emitted/absorbed) electromagnetic radiation, only discrete frequencies occur

$$\nu_{nm} = (E_n - E_m)/h.$$

This explains the atomic spectra.

As a final comment, it is worthwhile to note that the hydrogen atom provides a very important example of the phenomenon by which quantum Hamiltonians are more regular than their classical counterparts. In fact, for the classical Coulomb two-body problem the energy is not bounded below and under a perturbation, like emission of electromagnetic radiation, the system collapses with the electron falling on the nucleus; in the quantum case the Hamiltonian is bounded below. For the three-body problem the singularities of the interaction potentials which prevent the control of the global existence of solutions in the classical case become somewhat harmless in the quantum case, where the global existence is fully under control.

5.6 Appendix G: Properties of the Runge-Lenz operator

For the convenience of the reader, we reproduce the derivation of the eqs. (5.5.8), (5.5.9) since it is usually omitted in the literature or presented in a perhaps less direct form. We start by computing $\sum_i A_i^2$. Since

$$A_i = x_i/r - \varepsilon_{ijk} p_j \, L_k + i \, p_i = x_i/r - \varepsilon_{ijk} \, L_k p_j - i \, p_i,$$

we have (summation over repeated indices understood)

$$A_i A_i = (x_i/r - \varepsilon_{ijk} \, L_k p_j - i \, p_i)(x_i/r - \varepsilon_{iln} p_l \, L_n + i \, p_i)$$

$$= 1 - \varepsilon_{ijk}(x_i/r)p_j L_k + i(x_i/r)p_i - \varepsilon_{ijk} L_k \, p_j(x_i/r) + \varepsilon_{ijk}\varepsilon_{iln} L_k p_j p_l L_n$$

$$-i\varepsilon_{ijk} L_k p_j p_i - ip_i(x_i/r) + i\varepsilon_{iln} p_i p_l L_n + p^2.$$

Since

$$i[(x_i/r), p_i] = -2/r, \quad \varepsilon_{ijk}\varepsilon_{iln} L_k p_j p_l L_n = L^2 p^2 - (\mathbf{L} \cdot \mathbf{p})^2 = L^2 \, p^2,$$

$$-\varepsilon_{ijk}((x_i/r)p_j L_k + L_k p_j(x_i/r)) = -r^{-1} L_k^2 - L_k^2 r^{-1} = -2L_k^2/r,$$

we get $\mathbf{A}^2 = 1 + H + H \, L^2$. Now we compute the commutator $[A_i, A_j]$. For this purpose, it is convenient to note that

$$A_j = x_j/r - 1/2\varepsilon_{jkl}(p_k L_l + L_l p_k)$$

$$= x_i/r - x_j p^2 - p_j \mathbf{x} \cdot \mathbf{p} = A_j^* = x_j/r - p^2 x_j - \mathbf{p} \cdot \mathbf{x} p_j,$$

where we have used

$$\mathbf{x} \cdot \mathbf{p} = \mathbf{p} \cdot \mathbf{x} + 3i, \quad [\mathbf{p} \cdot \mathbf{x}, p_j] = ip_j = [\mathbf{x} \cdot \mathbf{p}, p_j].$$

Now

$$\varepsilon_{ijk}[A_j, A_k] = 2\varepsilon_{ijk} A_j A_k$$

$$= \varepsilon_{ijk}(x_j/r - p^2 x_j - \mathbf{p} \cdot \mathbf{x} p_j)(x_k/r - x_k p^2 - p_k \mathbf{x} \cdot \mathbf{p})$$

$$= -2 \, L_i(p^2 \mathbf{x} \cdot \mathbf{p} - \mathbf{p} \cdot \mathbf{x} p^2 + [x_k/r, p_k]) = -2i \, L_i(p^2 - 2/r) = -2i \, H L_i$$

and by multiplying the so obtained equation by ε_{iln} and summing over i, one gets eq. (5.5.9).

Chapter 6

* Quantum mechanics and stochastic processes

6.1 Quantum mechanics, probability and diffusion

As stressed before, the relation between quantum mechanics and probability is very strong. Indeed, for any commutative C^*-algebra generated by a set of commuting observables, a state defines a probability measure on its spectrum; in particular, if $\mathcal{A}_q$ denotes the C^*-algebra generated by the Weyl operators $U(\alpha) = \exp(i\alpha q)$, $\alpha \in \mathbf{R}$, then the Fock state (see Sect. 3.4) defines a Gaussian probability measure on the spectrum $\mathbf{R}$ of q so that q becomes a Gaussian random variable with zero mean and variance $1/2$.

A deeper relation between quantum mechanics and probability theory arises from the following relation: the propagator of a free particle in the x representation, namely the kernel of $\exp(-iH_0(t - t_0))$ as integral operator in $L^2(\mathbf{R}, dx)$, (we consider the one dimensional case and use units such that $\hbar = 1 = m$)

$$G_0(x, x_0; t - t_0) = \begin{cases} (2\pi i(t - t_0))^{-1/2} \, e^{-(x-x_0)^2/2i(t-t_0)}, & (t - t_0) > 0, \\ \delta(x - x_0), & (t - t_0) = 0, \end{cases}$$

becomes the kernel K of the semigroup $\exp(-H_0\tau)$, $\tau \in \mathbf{R}^+$, for the heat or diffusion equation

$$\partial u / \partial \tau = D\Delta u, \tag{6.1.1}$$

$$K(x, x_0; \tau - \tau_0) = \begin{cases} (4\pi D(\tau - \tau_0))^{-1/2} \, e^{-(x-x_0)^2/4D(\tau-\tau_0)}, & \tau - \tau_0 > 0, \\ \delta(x - x_0), & \tau - \tau_0 = 0, \end{cases}$$

with diffusion constant $D = 1/2$, [1] provided one takes $t = -i\tau$, $\tau \in \mathbf{R}^+$.

This relation is sometimes briefly summarized by the catch-words that for imaginary time the Schroedinger equation goes into the heat or diffusion equation (Sects. 6.3, 6.5).

Now, as discovered by Einstein, [2] an important property of the diffusion equation is its probabilistic interpretation in terms of a stochastic process, the Brownian motion (see below). Einstein's discovery opened a new way of looking to diffusion-like equations and stimulated Wiener in laying the foundations of the mathematical theory of stochastic processes [3] and their relation with parabolic equations. [4] In view of the above relation between the Schroedinger and the diffusion equation, a natural question is whether also quantum mechanics has a similar interpretation in terms of classical stochastic processes. That this is indeed the case has been advocated by Feynman, [5] who suggested an interpretation of quantum mechanics, in particular of the Schroedinger propagator e^{-iHt}, in terms of "averages" over classical paths or trajectories, (*Feynman path integral*) and later taken up by Nelson. [6] A derivation will be presented in Sect. 6.2.

A full discussion of this problem falls outside the scope of these notes; in the following sections we will address the following basic questions:
1) in which sense can the Feynman path integral approach lead to a stochastic interpretation of quantum mechanics? How far and how deeply can such a stochastic interpretation of quantum mechanics be carried through?
2) Apart from the philosophical implications of a stochastic interpretation of quantum mechanics, what is the advantage of looking at quantum mechanics from the stochastic processes (or functional integral) point of view?

As we shall see, whereas the link between imaginary-time quantum mechanics and stochastic processes is mathematically sound, the Feynman path integral for e^{-iHt} is more problematic,[7] since it is not an integral in the conventional mathematical sense, i.e. it is not defined by a σ-additive

---

[1] The heat propagator K can be easily derived as the solution of the Fourier transform of the heat equation with initial condition $K = \delta(x - x_0)$.

[2] A. Einstein, *Investigations on the theory of brownian movement*, Dover 1956.

[3] For a historical review see M. Kac, Bull. Amer. Math. Soc. **72**, Part II, 52 (1966).

[4] Actually, quite generally, the contraction semigroups which describe the time evolution of parabolic equations can be expressed in terms of probability measures (see, e.g. D.W. Stroock and S.R.S. Varadhan, *Multidimensional Stochastic Processes*, Springer 1979, esp. Chap. 3) so that such equations have a physical interpretation in terms of statistical or probabilistic phenomena.

[5] R. Feynman, Rev. Mod. Phys. **20**, 367 (1948).

[6] E. Nelson, Jour. Math. Phys. **5**, 332 (1964); *Dynamical theories of Brownian motion*, Princeton Univ. Press 1967; *Quantum Fluctuations*, Princeton Univ. Press 1985.

[7] For a discussion see S. Albeverio and R. Høegh-Krohn, *Mathematical theory of Feynman path integrals*, Springer 1976; S.Albeverio, S. Paycha and S. Scarlatti, in *Functional Integration, Geometry and Strings*, Z. Haba and J. Sobczyk eds., Birkhäuser 1989; L.S. Schulman, *Techniques and Applications of Path Integration*, J.Wiley 1981.

measure, as in the ordinary case[8] (see below and Appendix K). Hence, the probabilistic interpretation of the Feynman path integral in terms of a conventional stochastic process seems to be precluded. On the other hand, the relation between imaginary-time quantum mechanics and stochastic processes is not only interesting from a conceptual point of view (Sects. 6.5, 6.6), but also because it yields non-trivial technical tools for solving or shedding light on important quantum mechanical problems, since the wisdom gained on stochastic processes can be exploited (Sects. 6.4, 6.7).[9]

To motivate the above considerations better, we briefly review the probabilistic interpretation of the heat equation, discovered by Einstein. For this purpose, following Smoluchowski approach, we discretize the space of the motion, for simplicity taken one-dimensional, and approximate the motion of the Brownian particle by a jump of length Δx during each discretized time spacing $\Delta \tau$. The probabilistic character of this motion is described by giving a probability p that the jump is forward (and $q = 1 - p$ that the jump is backward)(*one-dimensional random walk*). Successive jumps are considered as independent and we denote by X_j the jump at time $j\Delta\tau$. For simplicity, take $p = q$. Then, the X_j are independent random variables with zero mean and variance

$$< X_j^2 >= 1/2(\Delta x)^2 + 1/2(\Delta x)^2 = (\Delta x)^2, \quad < X_j X_k >= \delta_{jk}(\Delta x)^2.$$

If $D(n)$ denotes the displacement at time $n\Delta\tau$, its probability distribution is given by

$$P(D(n) = j\Delta x) = \frac{n!}{((j+n)/2)!\,((n-j)/2)!}\frac{1}{2^n} = \binom{n}{(j+n)/2}\frac{1}{2^n}$$

for $j + n$ even and zero otherwise, since the number $n_{forw/back}$ of forward/backward jumps satisfy $n_{forw} - n_{back} = j$, $n_{forw} + n_{back} = n$, i.e. $n_{forw} = (n+j)/2$.

To discuss the space and time continuum limit, it is convenient to define the displacement at time τ by linearly interpolating the displacement between $[\tau/\Delta\tau]$ and $[\tau/\Delta\tau] + 1$, where the square brackets denote the integer part. Then, the mean square displacement is given by

$$< D_n(\tau)^2 >= (\Delta x)^2 [\tau/\Delta\tau], \quad D_n(\tau) = \sum_{j=1}^{[\tau/\Delta\tau]} X_j, \ n = [\tau/\Delta\tau].$$

Since, in the continuum limit given by $\Delta x \to 0$, $\Delta\tau \to 0$, we aim to describe a particle motion with non-zero mean square displacement, $(\Delta x)^2/\Delta\tau$ must

---

[8]R.H. Cameron, Jour. Math. Phys. **39**, 126 (1960).

[9]See the comprehensive book by B. Simon, *Functional Integration and Quantum Physics*, Academic Press 1979 and M. Kac, *Integration in Function Spaces and Some of Its Applications*, SNS Pisa 1980; S. Albeverio, in *Proceedings of Symposia in Applied Mathematics Vol.52*, Am. Math. Soc. 1997, p. 163.

converge to a non-zero limit, say α, and we have to take $\Delta x \simeq \alpha \sqrt{\Delta \tau}$. The probability distribution of the displacement $D_n(\tau)$ at time $\tau = n\Delta\tau$ is the same as that of the normalized sum

$$S_n(\tau) \equiv \frac{1}{\sqrt{n}} \, (\sum_{j=1}^{n} Y_j),$$

where the $Y_j \equiv \sqrt{n} X_j$ are identically distributed random variables with zero mean and variance $n\alpha^2 \Delta \tau = \alpha^2 \tau$.

For each fixed time τ, $\Delta \tau = \tau/n$ and the continuum limit corresponds to letting $n \to \infty$. Then, by the central limit theorem[10] the probability distribution of $S_n(\tau)$, and that of $D_n(\tau)$, converges weakly to the Gaussian distribution

$$\rho(x,\tau) \, dx = (2\alpha^2 \, \tau \, \pi)^{-1/2} e^{-x^2/2\alpha^2\tau} \, dx.$$

One easily recognizes that $\rho(x,\tau)$ is the propagator for the heat or diffusion equation, (with diffusion constant $D = \alpha^2/2$), and therefore the latter can be interpreted as the probability that a particle moving with a random walk and initially in zero (i.e. with probability distribution $\delta(x)$) be in x at time τ; hence, the solution $u(x,\tau)$ of the heat equation describes the random walk propagation of the initial probability distribution $u(x,0)$. The heat propagation can then be seen as the result of random steps, in the limit in which the microscopic size of the steps is negligible with respect to the macroscopic scale of space and of time (*continuum limit*), i.e. the heat propagation is a probabilistic propagation . Equivalently, as discovered by Einstein, the probability distribution density of the random Brownian motion is the fundamental solution of the diffusion equation. [11]

The connection between the random walk and the diffusion equation can also be seen by noticing that, putting $\Delta \equiv \Delta x, \delta \equiv \Delta \tau, p(n\Delta, (k+1)\delta) \equiv P(D((k+1)\delta) = n\Delta)$, one has

$$p(n\Delta, (k+1)\delta) = (1/2) \, p((n+1)\Delta, k\delta) + (1/2) \, p((n-1)\Delta, k\delta),$$

i.e.

$$[p(n\Delta, (k+1)\delta) - p(n\Delta, k\delta)]/\delta$$
$$= (\Delta^2/2\delta)[p((n+1)\Delta, k\delta) - 2p(n\Delta, k\delta) + p((n-1)\Delta, k\delta)]/\Delta^2.$$

In the r.h.s. one recognizes the finite difference approximation of the Laplacean, so that, letting $\delta, \Delta \to 0$, in such a way that $\Delta^2/2\delta = D$, $n\Delta \to x$, $k\delta = \tau$, the above difference equation goes into the diffusion equation (6.1.1) for the probability distribution density $p(x,\tau)$.

---

[10]See e.g. J. Lamperti, *Probability*, Benjamin 1966; a short proof is given in Appendix H, for the convenience of the reader.

[11]The continuum limit of the asymmetric random walk, corresponding to $p \neq q$, gives the probability distribution function of a Brownian particle in the presence of a constant force; see e.g. M. Kac, Am. Math. Monthly, **54**, 369 (1954).

6.2   The Feynman path integral

The relation between quantum mechanics and (lagrangean) classical mechanics was obviously a question of interest for the founders of quantum mechanics and in a seminal paper [12] Dirac proposed an analogy between the quantum propagator, i.e. the kernel of $\exp(-iH(t - t_0)/\hbar)$, and the exponential of the classical action, i.e. the integral of the classical Lagrangean, and argued that in the limit $\hbar \to 0$

$$G(x, x_0; t - t_0) \approx N e^{i \, S_{t - t_0}(x, x_0)/\hbar},$$

where N is a normalization constant and $S_{t - t_0}(x, x_0)$ is the classical action for trajectories starting from x_0 and ending at x, during the time interval $t - t_0$.

Indeed a simple relation can be established in the case of a free particle: (for simplicity we consider the one-dimensional case) the classical trajectory is given by $x(s) = x_0 + s\,v$, $s \in [t, t_0]$, so that

$$S_{t, t_0}(x, x_0) = (m/2) \int_{t_0}^{t} ds (dx(s)/ds)^2 = (m/2)\, v^2\, (t - t_0)$$

$$= m(x - x_0)^2 / 2(t - t_0),$$

since $v = (x(t) - x_0)/(t - t_0)$. Thus one has

$$G_0(x, x_0; t, t_0) = \left(\frac{m}{2\pi i(t - t_0)\hbar} \right)^{1/2} e^{iS_{t, t_0}(x, x_0)/\hbar}. \tag{6.2.1}$$

The non-trivial extension of such a relation to the case of non-zero potential was the great achievement by Feynman [13] and led him to the discovery of the Feynman path-integral representation of e^{-iHt}.

The following is one of the standard derivations of the *Feynman path-integral representation* of e^{-iHt} [14] as a limit of averages over polygonal paths when the time difference of the corners of the polygon tends to zero.

---

[12] P.A.M. Dirac, Phys. Zeit. Sowietunion, **3**, 64, 1933, reprinted in *Selected Papers on Quantum Electrodynamics*, J. Schwinger ed., Dover 1958; see also P.A.M. Dirac, *The Principles of Quantum Mechanics*, Oxford Claredon Press 1958, Sect. 32.

[13] R.P. Feynman, Rev. Mod. Phys. **20**, 367 (1948); see also R.P. Feynman and A.R. Hibbs, *Quantum Mechanics and Path Integrals*, McGraw-Hill, 1965. For the history and the anecdotes see S. Schweber, *QED and the men who made it*, Princeton Univ. Press 1994, Chap. 8.

[14] See M. Reed and B. Simon, *Methods of Modern Mathematical Physics, Vol. II*, Academic Press 1975; L.S. Schulman, *Techniques and Applications of Path Integration*, J. Wiley 1981; S. Albeverio, in *Proceedings of Symposia in Applied Mathematics Vol. 52*, V. Mandrekar and P.R. Masani eds., Am. Math. Soc. 1997, p. 163 and references therein.

The starting point is Trotter product formula which generalizes the classical theorem by Lie (Lie product formula) according to which: if A and B are finite-dimensional matrices then

$$e^{A+B} = \lim_{n \to \infty} (e^{A/n} e^{B/n})^n.$$

Theorem 6.2.1 *(Trotter product formula) If H_0 and V are self-adjoint operators and $H_0 + V$ is essentially self-adjoint on the dense domain $D = D(H_0) \cap D(V)$, then*

$$e^{-it(H_0+V)} = strong - \lim_{n \to \infty} (e^{-itH_0/n} e^{-itV/n})^n. \tag{6.2.2}$$

Moreover, if H_0 and V are bounded below, the same formula holds for imaginary times $t = -i\tau$, $\tau \in \mathbf{R}^+$.

Proof. We briefly sketch the proof for H_0, V bounded. [15] Putting

$$C_n \equiv e^{(A+B)/n}, \quad D_n \equiv e^{A/n} e^{B/n}$$

we have to prove that $||D_n^n - C_n^n|| \to 0$ and for this we exploit the identity

$$D^n - C^n = \sum_{k=1}^{n} D^{k-1} (D - C) C^{n-k}.$$

Since

$$||C_n|| \leq e^{(||A||+||B||)/n}, \quad ||D_n|| \leq e^{||A||/n} e^{||B||/n},$$

we have

$$||D_n^n - C_n^n|| \leq \sum_{k=1}^{n} e^{(||A||+||B||)(k-1)/n} ||D_n - C_n|| e^{(||A||+||B||)(n-k)/n}$$

$$\leq n ||D_n - C_n|| e^{(||A||+||B||)}.$$

Now, by expanding the exponentials we have

$$D_n - C_n = R/n^2, \quad R = [A, B]/2 + O(1/n)$$

and

$$||D_n^n - C_n^n|| = O(1/n) \to 0.$$

---

[15] For the proof when $H_0 + V$ is self-adjoint see E. Nelson, Jour. Math. Phys. **5**, 332 (1963); M. Reed and B. Simon, *Methods of Modern Mathematical Physics, Vol.I*, Academic Press 1972, Sect. VIII.8; B. Simon, *Functional Integration and Quantum Physics*, Academic Press 1979, p. 4. For the case of essential self-adjointness and imaginary times $t = -i\tau$, $\tau \in \mathbf{R}^+$, see J. Glimm and A. Jaffe, *Quantum Physics. A Functional Integral Point of View*, Springer 1987, Appendix A.5.

The idea of the proof can be extended to the case of $H_0 + V$ self-adjoint (see Nelson, loc. cit.). [16]

We can now derive *the Feynman path formula*. For simplicity, we consider the one dimensional case, with V continuous and bounded below.

By using eq. (6.2.1), for the kernel of $\exp(-itH_0/\hbar)$, $\forall \psi \in \mathcal{S}(\mathbf{R})$ we have

$$(e^{-iH_0 t/n\hbar} e^{-iVt/n\hbar} \psi)(x_1)$$

$$= N_n \int e^{im(x_1 - x_0)^2 / 2\hbar(t/n)} e^{-iV(x_0)t/n\hbar} \psi(x_0) dx_0,$$

where $N_n \equiv (nm/2\pi i\hbar t)^{1/2}$, and by applying Trotter formula to the Hamiltonian $H = H_0 + V$, we get

$$(e^{-iHt/\hbar} \psi)(x) = \lim_{n\to\infty} N_n^n \int dx_{n-1}...dx_1 e^{iS_{t,n}(x,x_{n-1},...,x_0)/\hbar} \psi(x_0) \, dx_0,$$

$$(6.2.3)$$

$$S_{t,n}(x, x_{n-1}, ..., x_0) \equiv \sum_{i=0}^{n-1} \frac{t}{n} \left[\frac{m}{2} \left(\frac{x_{i+1} - x_i}{t/n} \right)^2 - V(x_i) \right], \quad x_n \equiv x,$$

where all improper integrals are defined as symmetric infinite volume limits and all limits are in the L^2 sense (furthermore the formula extends to all $\psi \in \mathcal{H}$, since $\exp -itH/\hbar$ and its Trotter approximants are bounded operators).

To visualize eq. (6.2.3) and get an interpretation of the limit, introduce the polygonal (piecewise linear) path $\gamma_n(s)$, $s \in [0, t]$, with $\gamma_n(t_j \equiv jt/n) = x_j$, which can be regarded as the n-th step approximation of a trajectory $\gamma(s)$ passing through the points $x_1, ... x_{n-1}$ at times $t_j = jt/n$. As n increases, one gets a finer time slicing and a more refined point crossing by the trajectory. Then, S_n can be seen as the Riemann sum approximation of the integral

$$S_t(\gamma(\cdot)) = \int_0^t ((m/2)\dot{\gamma}(s)^2 - V(\gamma(s)))ds.$$

---

[16] One uses the same identity on $\psi \in D$:

$$||(D_n^n - C_n^n)\psi|| \leq \sup_{0 \leq s \leq t} n\,||(D_n - C_n)\psi_s||, \quad \psi_s \equiv e^{-is(H_0+V)}\psi.$$

Furthermore, for each s, $\lim_n n\,||(D_n - C_n)\psi_s|| = 0$, since for $\varepsilon = t/n \to 0$

$$s - lim_{\varepsilon \to 0} \varepsilon^{-1}(e^{-i\varepsilon A} e^{-i\varepsilon B} - e^{-i\varepsilon(A+B)})\psi_s = 0.$$

A uniform estimate for $s \in [0, t]$ is obtained by the uniform boundedness principle applied to D equipped with the norm $|||\chi||| \equiv ||(A + B)\chi|| + ||\chi||$, which makes D a Banach space and $n(D_n - C_n)$ a family of bounded operators, so that

$$n\,||(D_n - C_n)\psi|| \leq \text{const } |||\psi|||.$$

Moreover, the set $\{\psi_s, s \in [0, t]\}$ is compact, since $[0, t] \ni s \to \psi_s \in D$ is $|||\;\;|||$-continuous; hence the above limit in n is uniform.

On the other side, the integral over the intermediate points $x_1, ..., x_{n-1}$ can be interpreted as the integration over all possible paths or trajectories starting from x_0 at $t = 0$ and ending at x at time t.

Thus, one is led to the Feynman path integral formula for the propagator

$$G(x, x_0; t, 0) = \lim_{n \to \infty} N_n^n \int dx_1 ... dx_{n-1} e^{iS_{t,n}(x, x_{n-1}, ..., x_0)/\hbar}$$

$$= \int_{\gamma(0)=x_0, \gamma(t)=x} \mathcal{D}\gamma(\cdot) \, e^{(i/\hbar)S_t(\gamma(\cdot))}, \tag{6.2.4}$$

where the first equation is a mathematical result and in the second $\mathcal{D}\gamma(\cdot)$ has the "heuristic" meaning of integration over all the classical trajectories satisfying the given boundary conditions at time 0 and t.

From a physical point of view, this formula has strong conceptual implications, since it suggests that the quantum propagation from x_0 to x in the time interval $[t_0, t]$ is the result of an "average" over the classical paths starting from x_0 at time t_0 and ending at x at time t, just as the propagation of a Brownian particle can be seen as the result of an average over classical paths (as discussed in Sect. 6.1 and in more detail in the following Section). It is clear that this point of view provides a completely new and revolutionary interpretation of quantum mechanics, since it establishes an unexpected link with classical mechanics and it suggests a strong relation with stochastic processes.

The formula (6.2.4) also provides a simple way of understanding the relation between classical and quantum mechanics. In fact, in the limit $\hbar \to 0$ by the stationary phase approximation the contributions from regions in path space in which $S_t(\gamma(\cdot)) \neq 0$ is not stationary are expected to be washed out by the oscillatory phase (at least of order $\hbar$); the dominant contribution comes from the stationary points of the classical action, i.e. from the classical solutions. The quantum effects, i.e. the deviations from the classical solutions, are then interpreted as "quantum" fluctuations (see Sect. 6.7.2 for more details).

From a practical point of view, such a formula has played a very important role in the development of quantum mechanics and has led to the breakthrough of Feynman perturbative expansion for Quantum Electrodynamics and to its successful predictions (for a historical perspective see the quoted book by Schweber). It should also be mentioned that the Feynman path integral formula has strongly influenced the development of the theory of stochastic processes.

From a technical point of view the first equality in eq. (6.2.4) has a sound mathematical basis and can be taken as the definition of the Feynman path-integral; the name "integral" merely accounts for its being the limit of integrals. The second equality, which writes the limit as a functional integral meets non-trivial mathematical problems. [17]

The symbol $\mathcal{D}\gamma(\cdot)$, which should be the limit of $N_n^n \prod_{j=1}^{n-1} dx_j$, is ill defined because both N_n and the product of flat Lebesgue measures are divergent when $n \to \infty$. Actually, as proved by Cameron

$$\mathcal{D}\gamma(\cdot)\, e^{i \int_0^t \dot{\gamma}(s)^2\, ds}$$

cannot define a measure on path space in the standard sense (see Appendix K) and therefore the Feynman path integral is not a (functional) integral; a probabilistic interpretation, though very suggestive, is thus precluded. Furthermore, for the existence of the classical action $\int_0^t \dot{\gamma}(s)^2\, ds$ the trajectories involved in the sum should have square integrable derivatives, but nothing prevents the relevant paths from being rather irregular (as we shall see below) so that for them the classical action is not defined [18]; thus the integrand is also meaningless.

As we shall see in the next Section, the fundamental contribution by Kac was to realize that such problems are overcome for imaginary times, i.e. for the semigroup $e^{-\tau H}$, $\tau \geq 0$, since then the divergence of the normalization constant compensates for the vanishing of $\exp - \int_0^\tau \dot{\gamma}(s)^2\, ds$. The result is that

$$N_n^n(\tau) dx_{n-1}...dx_1 \, \exp - \sum_{i=0}^{n-1} \frac{\tau}{n} \left[\frac{m}{2} \left(\frac{x_{i+1} - x_i}{\tau/n} \right)^2 \right]$$

converges to a well defined measure in the limit $n \to \infty$, the *Wiener measure* for the Brownian motion, and provides an explicit construction of a measure in an infinite dimensional space. The corresponding theory is called "euclidean quantum mechanics" because in quantum field theory the imaginary time turns Minkowski space into a euclidean space.

---

[17] For a general overview of the mathematical problems of the Feynman path-integral see S. Albeverio, S. Paycha and S. Scarlatti, in *Functional Integration, Geometry and Strings*, Proceedings of the XXV Karpacz Winter School, Z. Haba and J. Sobczyk eds., Birkhäuser 1989, p. 230, which contains an extensive bibliography.

[18] For a simple check of the fact that the set of paths with finite action are irrelevant, even in the more regular euclidean case, see S. Coleman, *Aspects of Symmetry*, Cambridge University Press 1985, Chap. 7, Appendix 3.

6.3   The Feynman-Kac formula

A full probabilistic interpretation of the Feynman path integral can be obtained if one takes imaginary times, i.e. if one considers the kernel of $e^{-\tau H}$, $\tau \geq 0$, rather than that of e^{itH}, as discovered by Kac, (*Feynman-Kac formula*). [19] The interest of such a formula is both technical and philosophical. First, the control of the semigroup $e^{-\tau H}$, $\tau \geq 0$, gives in any case non-trivial information on the unitary group e^{itH}; in particular one may learn about non-perturbative properties of the spectrum of H and of its eigenfunctions. [20]

Secondly, and more importantly, (as we shall discuss in Sect. 6.5), if the Hamiltonian has a lowest energy state (ground state) Ψ_0, the complete description of a system of quantum mechanical particles (including the solution of the dynamical problem) is fully encoded (see Sect. 6.5) in the so-called correlation functions (Wightman functions)

$$\mathcal{W}(t_1, ..., t_n) = (\Psi_0, q(t_1)...q(t_n)\, \Psi_0).$$

Now, as a consequence of the positivity of the energy with respect to the ground state, such functions are boundary values of analytic functions, with analyticity domains which contain the imaginary time points $\tau_1 = it_1, ..., \tau_n = it_n$ (see Sect. 6.5 below). Thus, the imaginary time correlation functions (called Schwinger functions) fully determine the theory; for them the Feynman-Kac formula applies and a probabilistic interpretation is possible. Also from a mathematical point of view the gain is very relevant since the standard measure theoretical wisdom becomes available. Thirdly, one has a mathematical control on *integration in infinite dimensional path space*.

We briefly sketch the proof of the Feynman-Kac formula, eq. (6.3.2) below, referring the interested reader to more complete accounts.[21] Our aim is to indicate the main ideas involved in the proof and to point out the crucial role of the imaginary time in providing the damping factor

$$e^{-\mathcal{T}_n} \equiv e^{-\sum_{i=1}^{n} \frac{\tau}{n}\frac{m}{2}\left(\frac{x_i - x_{i-1}}{\tau/n}\right)^2}$$

which, when combined with the proper normalization constant, leads to the Wiener measure in the limit $n \to \infty$, and to the link with stochastic processes and with integration in infinite dimensional spaces.

---

[19] M. Kac, *Proc. 2nd Berkeley Symp. Math. Stat. Probability*, (1950), p. 189.

[20] See B. Simon, *Functional Integration and Quantum Mechanics*, Academic Press 1979, for an extensive discussion, as well as S. Albeverio, in *Proc. Symp. Appl. Math. Vol. 52*, Am. Math. Soc. 1997, p. 163.

[21] See, e.g. J. Glimm and A. Jaffe, *Quantum Physics. A Functional Integral Point of View*, Springer 1987, Sect. 3.2; M. Reed and B. Simon, *Methods of Modern Mathematical Physics. Vol. II*, Academic Press 1975, Sect. X.11.

For simplicity, we shall assume that the potential V is (real) continuous and bounded from below.

The starting point is again the Trotter formula leading to the analog of eq. (6.2.3), with it replaced by τ, (for simplicity we put $m = 1, \hbar = 1$),

$$(e^{-\tau H}\psi)(x) = \lim_{n \to \infty} (N_n^E)^n \int dx_{n-1}...dx_1 \, e^{-S_n^E(x, x_{n-1},...,x_0)}\psi(x_0) \, dx_0,$$

$$S_n^E(x, x_{n-1}, ..., x_0) \equiv \sum_{i=0}^{n-1} \frac{\tau}{n} \left[\frac{1}{2} \left(\frac{x_{i+1} - x_i}{\tau/n} \right)^2 + V(x_i) \right], \quad x_n \equiv x,$$

with $N_n^E \equiv (n/2\pi\tau)^{1/2}, \psi \in \mathcal{S}$.

To control the limit on the right hand side, we note that for n operators $A_j, j = 1, ..., n$, in $L^2(\mathbf{R}, dx)$, acting as multiplication by bounded continuous functions $A_j(x)$, we have ($\tau_j \equiv j\tau/n$)

$$(e^{-(\tau/n)H_0} A_1 \, e^{-(\tau/n)H_0} A_2 ... e^{-(\tau/n)H_0} A_n \, \psi)(x_0)$$

$$= \int dx_1 ... dx_n K(x_0, x_1; \tau_1) A_1(x_1) ... K(x_{n-1}, x_n; \tau_n - \tau_{n-1}) A_n(x_n) \, \psi(x_n)$$

$$\equiv \int dx_1 ... dx_n \, P_{\tau_1, ..., \tau_n}(x_0, x_1, ..., x_n) \prod_{j=1}^{n} A_j(x_j) \, \psi(x_n). \qquad (6.3.1)$$

Where, thanks to the positivity of the kernel K of $e^{-\tau H_0}$, the P's have the meaning of joint probability distributions for functions which depend only on a finite number of points of the trajectories which start from x_0 at $\tau = 0$ and pass through $x_j \equiv x(\tau_j)$ at $\tau = \tau_j \equiv j\tau/n$. Such joint probabilities satisfy the compatibility conditions 1-3 (see Appendix J), as a consequence of the semigroup property of $e^{-\tau H_0}$. Thus, by the Nelson version of Kolmogorov theorem (see Appendix J), they define a functional measure on path space, i.e. the right hand side can be written as a functional integral in terms of the Wiener measure $dW_{x_0}(x(\cdot))$ over the space X of continuous trajectories starting from x_0 (see Appendix K)

$$\int_X dW_{x_0}(x(\cdot)) \prod_{j=1}^{n} A_j(x(\tau_j)) \, \psi(x_n), \quad x_n = x(\tau). \qquad (6.3.2)$$

To complete the proof of the Feynman-Kac formula we have to discuss the limit $n \to \infty$ of the above expression with $A_j = e^{-(\tau_j - \tau_{j-1})V(x_j)}$, $j = 1, ...n, x_j = x(j\tau/n), \tau_0 \equiv 0$. Now, for continuous paths, $V(x(s))$ is continuous and Riemann integrable in s; then, by the convergence of the Riemann sums to the integral, for each continuous path,

$$\lim_{n \to \infty} e^{-\sum_{j=1}^{n} (\tau_j - \tau_{j-1}) V(x(\tau_j))} = e^{-\int_0^\tau V(x(s)) \, ds} \equiv e^{-\mathcal{V}},$$

i.e. pointwise on the points of X corresponding to continuous paths. Since dW_{x_0}-almost all paths are continuous (Appendix K) the convergence is dW_{x_0}-almost everywhere in X. Moreover, since V is bounded from below $(\tau_j - \tau_{j-1} = \tau/n)$

$$\int_X |\psi(x(\tau)) e^{-\sum_{j=1}^n (\tau_j - \tau_{j-1}) V(x(\tau_j))}| \, dW_{x_0} \le e^{-\tau V_{min}} \int_X |\psi(x(\tau))| \, dW_{x_0}$$

$$= e^{-\tau V_{min}} (e^{-\tau H_0} |\psi|)(x_0) < \infty.$$

Thus, by the dominated convergence theorem $e^{-\int_0^\tau V(x(s)) \, ds}$ is dW_{x_0} integrable and the expression (6.3.1) converges to the Feynman-Kac formula

$$(e^{-\tau H} \psi)(x) = \int_X dW_x(x(\cdot)) \, e^{-\int_0^\tau V(x(s)) \, ds} \, \psi(x(\tau)). \tag{6.3.3}$$

The extension to $\psi \in L^2$ is obtained as a limit, by a dominated convergence argument (see the above estimate) and by the continuity of the operator $e^{-\tau H}$.

The above formula remains valid in the general case of $V \in L^2 + L^\infty$, where one gets the integrability of the potential $V(x(s))$ for dW_{x_0}-almost all trajectories (for the proof see Reed and Simon book quoted above).

The Feynman-Kac formula expresses the action of $e^{-\tau H}$ on an L^2 function as a functional integral over paths with a measure $d\mu_x \equiv dW_x e^{-V}$, which can be interpreted as a perturbation of the Wiener measure and it is the analog of the Feynman path integral weight $\mathcal{D}(x(\cdot)) \exp i S_t(x(\cdot))$, (which, instead, cannot be given the meaning of a measure in the standard mathematical sense).

The mathematical control of the limit yielding the measure $d\mu_x$, on the infinite dimensional path space, has been made possible by i) the positivity of the heat kernel, which provides the link with probability and measure theory, ii) the damping effect of the imaginary time, which yields both the convergence of $(N_n^E)^n \, e^{-T_n}$ and the integrability of e^{-V}. Both properties i) and ii) fail for real times and this failure is at the origin of the obstructions pointed out by Cameron for a measure theoretical content of the Feynman path integral.

6.4   Nelson positivity and uniqueness of the ground state

The Feynman-Kac (FK) formula does not only represent a conceptually deep result for its stochastic interpretation of the kernel $\mathcal{K}$ of $e^{-\tau H}$, but it also proves useful for global non-perturbative control of the eigenfunctions of H.

An important property which follows directly from the FK formula is the strict positivity of the kernel $\mathcal{K}$, if the potential is continuous and bounded below. In fact, one version of the FK formula expresses directly $\mathcal{K}$ as a functional integral

$$\mathcal{K}(x, x'; 0, \tau) = \int dW_{x,x'}(x(\cdot)) \, e^{-\int_0^\tau V(x(s)) \, ds}, \qquad (6.4.1)$$

where $dW_{x,x'}$ is the conditional Wiener measure, i.e. the Wiener measure with the condition that both end points of the path are fixed $x(\tau) = x', x(0) = x$:

$$dW_{x,x'} = dW_x \, \delta(x(\tau) - x').$$

The strict positivity of $\mathcal{K}$ easily follows from the above functional integral representation in terms of a positive measure.

Such a property has very important consequences and it leads to Nelson positivity of the Schwinger functions (see below). As pointed out by Glimm and Jaffe [22] it implies the uniqueness of the ground state and that its wave function can be chosen to be strictly positive.

The proof exploits a generalization of the Perron-Frobenius theorem, which states that the maximum eigenvalue of a matrix with strictly positive elements has no multiplicity and the corresponding eigenvector can be chosen positive. The generalization to a bounded symmetric operator A in $L^2(\mathbf{R}^s, dx)$ requires the *strict positivity of the kernel* $A(x,y)$, namely that for any (non-trivial) non-negative L^2-function ψ, $A\psi$ is strictly positive almost everywhere. In fact, this implies that A is a positive operator and if $\lambda \equiv \sup \sigma(A)$ is an eigenvalue of A, and ψ is a corresponding eigenfunction,

$$0 < \lambda = \int dx \, dy \, A(x, y) \, \overline{\psi(x)} \, \psi(y) \le \int dx \, dy \, A(x, y) |\psi(y)| \, |\psi(x)|.$$

Since $\lambda = \sup_\varphi |(\varphi, \, A \, \varphi)|$, the equality must hold. Hence, the function $A(x, y) \, \overline{\psi(x)} \, \psi(y)$ must be real and positive, since its integral coincides with that of its modulus. This implies

$$\psi(x) = e^{i\alpha} |\psi(x)|, \quad a.e., \quad \alpha \in \mathbf{R},$$

---

[22] See J. Glimm and A. Jaffe, *Quantum Physics. A Functional Integral Point of view*, Springer 1987, Sect. 3.3.

so that ψ can be chosen to be positive, actually strictly positive, since $\lambda \psi = A \psi > 0$, by the strict positivity of A.

Finally, if there were two eigenfunctions with eigenvalue λ, both could be chosen strictly positive and therefore could not be orthogonal.

In the case of the kernel $\mathcal{K}$, $\lambda \equiv \sup \sigma(e^{-H}) = e^{-\inf \sigma(H)}$ and if λ is an eigenvalue [23] the corresponding eigenfunction $\Psi_0(x)$ describes a ground state of H; by the strict positivity of $\mathcal{K}$ the ground state is therefore unique, up to a phase, and its wave function can be chosen strictly positive. The uniqueness of the ground state and the strict positivity of its wave function have strong mathematical and physical implications and it is very rewarding that they appear as simple consequences of the Feynman-Kac formula.

6.5 Quantum mechanics and stochastic processes

The links between quantum mechanics and stochastic processes can be strengthened by discussing the relation between (real time) quantum mechanics and its imaginary time version. We consider quantum systems with Hamiltonians satisfying the *stability condition*:

I.(Spectral condition) The spectrum of the Hamiltonian is bounded below: $\text{Inf } \sigma(H) = c > -\infty$.

This condition prevents the system from collapsing to lower and lower energy states under small (external) perturbations. It is this property which prevents the atoms from collapsing (see Sect. 5.5). The constant c can be put equal to zero, by a rescaling which does not change the dynamics. Actually, we shall consider systems for which the stability condition holds in the following stronger form [24]

I'.(Strong spectral condition) $0 = \text{Inf } \sigma(H)$ is an eigenvalue.

The corresponding eigenstate, which is unique (up to a phase) by the argument of Sect. 6.4, is called the *ground state* and denoted by Ψ_0.

For simplicity, we consider the case of a spinless particle subject to a potential, which is continuous and bounded below. As in the classical case, the solution of the dynamical problem is given by the knowledge of the time evolution of q, p. Actually, since $p(t) = m \dot{q}(t)$, it is enough to know the operators $q(t)$, $t \in \mathbf{R}$, equivalently $U(\alpha, t) \equiv \exp i\alpha q(t)$, $\alpha, t \in \mathbf{R}$, on a dense set. Now, the algebra $\mathcal{A}_{q,H}$ generated by the $U(\alpha, t)$'s is irreducible if

---

[23] For conditions which assure this property see M. Reed and B. Simon, *Methods of Modern Mathematical Physics*, Vol. IV, Academic Press 1978, Sect. XIII.12.

[24] In some cases, like for a periodic potential, this may require a non-regular representation of the canonical algebra; see e.g. J. Löffelholz, G. Morchio and F. Strocchi, Lett. Math. Phys. **35**, 251 (1995).

$\mathcal{A}_W$ is so, since if a bounded operator C commutes with $\mathcal{A}_{q,H}$, it commutes with $q(t)$ and $p(t)$, i.e. with $\mathcal{A}_W$. Hence, any vector, in particular Ψ_0, is cyclic for $\mathcal{A}_{q,H}$ and the ground state expectations (*Wightman functions*)

$$W_{\alpha_1,...,\alpha_n}(t_1,...,t_n) \equiv (\Psi_0,\, U(\alpha_1,t_1)...U(\alpha_n,t_n)\,\Psi_0) \qquad (6.5.1)$$

fully determine the operators $U(\alpha,\, t)$.

Under general regularity conditions for the dynamics[25] one may equivalently consider the ground state correlation functions of $q(t)$

$$W(t_1,...,t_n) = (\Psi_0, q(t_1)...q(t_n)\,\Psi_0) \equiv\; <q(t_1),...,q(t_n)>. \qquad (6.5.2)$$

Since the Wightman functions fully determine the theory it is important to point out their general properties. The invariance of Ψ_0 under time evolution implies the invariance of the correlation functions under time translations

W1. (*Time translation invariance*)

$$W_{\alpha_1,...,\alpha_n}(t_1,...,t_n) = W_{\alpha_1,...,\alpha_n}(t_1+s,...,t_n+s) \qquad (6.5.3)$$

$$\equiv W_{\alpha_1,...,\alpha_n}(\xi_1,...,\xi_{n-1}), \quad \xi_j \equiv t_{j+1}-t_j,$$

i.e. the W are actually functions of the difference variables.

For simplicity, by using a multiindex and multicomponent notation we shall denote the correlation functions by $W_\alpha(\xi)$ and sometimes even omit the multiindex α.

The strong continuity of the time evolution implies that the W's are continuous bounded functions of their arguments and therefore

W2. (*Temperedness*) The Wightman functions are tempered distributions.

By the spectral theorem, the spectral condition says that the Fourier transform $\tilde{W}(\omega)$ of $W(\xi)$ satisfies the following

W3. (*Spectral condition*)

$$\tilde{W}_\alpha(\omega) = 0, \;\; \text{if, for some } j, \;\; \omega_j < 0. \qquad (6.5.4)$$

This is easily seen e.g. for the two point function for which one has $\tilde{W}(\omega) = <q\,dE(\omega)\,q>$, with $dE(\omega)$ the spectral measure of the Hamiltonian and $dE(\omega) = 0$ for $\omega < 0$. Quite generally, $\forall\,\Psi,\,\Phi$,

$$F(t) \equiv (\Psi,\, e^{itH}\,\Phi) = \int (\Psi,\, dE(\omega)\,\Phi)e^{it\omega} \equiv \int \tilde{F}(\omega)d\omega\, e^{it\omega},$$

---

[25] E. Nelson, J. Funct. Anal. **11**, 211 (1972); J. Fröhlich, Comm. Math. Phys. **54**, 135 (1977). The conditions essentially amount to bound $\dot{q}(t)$, as an operator valued tempered distribution in t, by the Hamiltonian $\pm\dot{q}(f) \leq c_f(H+c)$, $f \in \mathcal{S}$, with c_f some continuous norm on $\mathcal{S}$ and c a constant. Then Ψ_0 is in the domain of the polynomials of q at various times and the corresponding ground state correlations are tempered distributions.

with supp $\tilde{F} \subseteq \mathbf{R}^+$. Thus the Laplace transform of $\tilde{F}$, $F(z = t + i\eta)$, $\eta > 0$, exists and it is analytic in the upper half plane $\eta = \text{Im } z > 0$.

Since the Wightman functions are defined by Hilbert scalar products one has the following [26]

W4. (*Hilbert Space Positivity condition*) For any terminating sequence $\{f_j \in \mathcal{S}(\mathbf{R}^j), j = 1...\}$, $f_0 \in \mathbf{C}$,

$$\sum_{j,k=0}^{N} \int dt_1...dt_{j+k} \, \bar{f}_j(t_j,...,t_1) \, \mathcal{W}_{\alpha_{jk}}(t_1,...,t_j; \, t_{j+1},...,t_k) \, f_k(t_{j+1},...,t_k) \geq 0.$$

$$\mathcal{W}_{\alpha_{jk}}(t_1,...,t_j; \, t_{j+1},...,t_k) \equiv \mathcal{W}_{-\alpha_1...-\alpha_j,\alpha_{j+1}...\alpha_k}(t_1,...,t_j,t_{j+1},...,t_k).$$

By a GNS type argument, one can easily show that from the ground state expectations of $\mathcal{A}_{q,H}$ one can reconstruct the Hilbert space and the quantum mechanical description of the system (Wightman reconstruction theorem), i.e. the Wightman functions encode the full (real time) quantum mechanical information.

The imaginary time version is obtained by noticing that the support properties of $\tilde{W}$ imply that its Laplace transform is an analytic function $W(\zeta) \equiv W(\zeta_1,...,\zeta_{n-1})$, for $\zeta_j = \xi_j + i\eta_j$, $\xi_j \in \mathbf{R}$, $\eta_j > 0$ and the original function $W(\xi)$ is the boundary value of $W(\zeta)$. Now, the analyticity domain of $W(\zeta)$ contains the imaginary time points, also called *euclidean points*, $\zeta = \{\zeta_j = i \, s_j, \, s_j > 0, \, j = 1,...,n-1\}$ and therefore the correlation functions at imaginary times (*Schwinger functions*) in the difference variables

$$S_\alpha(s) \equiv S_\alpha(s_1,...,s_{n-1}) \equiv W_\alpha(is_1,...,is_{n-1}) \tag{6.5.5}$$

determine the coefficients of the power series expansion of $W(\zeta)$ and therefore the theory at real time.

Clearly, one may also introduce imaginary time analogs of the $\mathcal{W}$

$$S_\alpha(\tau_1,...,\tau_n) \equiv W_\alpha(i\tau_1,...,i\tau_n) = S_\alpha(s_1,...,s_{n-1}), \quad s_j \equiv \tau_{j+1} - \tau_j. \tag{6.5.6}$$

The condition $s_j > 0$ means that the imaginary times are chronologically ordered: $\tau_1 < \tau_2 < ... < \tau_n$.

The Schwinger functions correspond to the following correlation functions

$$S_\alpha(s_1,...,s_{n-1}) = (\Psi_0, \, e^{i\alpha_1 q} \, e^{-s_1 H} \, e^{i\alpha_2 q}...e^{-s_{n-1} H} \, e^{i\alpha_n q} \, \Psi_0), \quad s_j > 0,$$

which involve the probabilistic kernel e^{-sH}. Thus, they can be obtained by a functional integral. For $-T < \tau_1 < \tau_2 < ... < \tau_n < T$, $s_j = \tau_{j+1} - \tau_j > 0$ one can write the above expectation value as

$$\int dx \, dx' \, \bar{\Psi}_0(x) \Psi_0(x') \int dW_{x,x'}(x(\cdot)) \, e^{-\int_{-T}^{T} V(x(s))ds} \, e^{i\alpha_1 x(\tau_1)}...e^{i\alpha_n x(\tau_n)},$$

---

[26] The condition follows from the positivity of the scalar products of vectors obtained by applying polynomials of the operators $e^{i\alpha q}(f) = \int dt \, f(t) \, U(\alpha, t)$ to Ψ_0.

where $dW_{x,x'}$ is the conditional Wiener measure for trajectories with end points $x(-T) = x'$, $x(T) = x$, see eq. (6.4.1) with a symmetric choice of the initial and final times $-T$, T.

The above expression involves the wave function of the ground state, whose knowledge is part of the solution of the dynamical problem. To get a more useful formula from a constructive point of view, one notices that by the spectral theorem and the strong spectral condition I', for any positive L^2 wave function Φ_0 one has

$$\Psi_0 \left(\Psi_0, \, \Phi_0\right) = \lim_{T \to \infty} e^{-2TH} \, \Phi_0,$$

$$Z \equiv |(\Psi_0, \, \Phi_0)| = \lim_{T \to \infty} ||e^{-TH}\Phi_0|| \equiv \lim_{T \to \infty} Z_T.$$

Therefore, the Schwinger functions can be obtained by the following limit

$$\lim_{T \to \infty} Z_T^{-2} \left(\Phi_0, \, e^{-(\tau_1+T)H} e^{i\alpha_1 q} e^{-s_1 H} e^{i\alpha_2 q} ... e^{-s_{n-1}H} e^{i\alpha_n q} e^{-(T-\tau_n)H}\Phi_0\right)$$

$$= \lim_{T \to \infty} Z_T^{-2} \int d\mu_W \left(x(\cdot)\right) e^{-\int_{-T}^{T} V(x(s))\,ds} \, e^{i\alpha_1 x(\tau_1)} ... e^{i\alpha_n x(\tau_n)}, \qquad (6.5.7)$$

where

$$d\mu_W \left(x(\cdot)\right) \equiv \int dx \, dx' \, \bar{\Phi}_0(x) \, \Phi_0(x') \, dW_{x,x'} \left(x(\cdot)\right)$$

and Φ_0 has the meaning of boundary conditions at times $-T$, T.

It is worthwhile to remark that the functional integral representation of the Schwinger functions has a meaning also for unordered (imaginary) times, and therefore one may define an extension of the Schwinger functions for arbitrary times. The so defined functions satisfy the following property

$$S_{\alpha_{\pi(1)} ... \alpha_{\pi(n)}} \left(\tau_{\pi(1)}, ..., \tau_{\pi(n)}\right) = S_{\alpha_1 ... \alpha_n} \left(\tau_1, ..., \tau_n\right),$$

where π denotes an arbitrary permutation of the indices $1, ..., n$. Correspondingly, the Schwinger functions S of the difference variables, eq. (6.5.6), get defined also for non positive arguments.

From a constructive point of view, it is convenient to consider the perturbation with respect to the harmonic oscillator rather than to the free particle (whose ground state cannot be described by an L^2 wave function). For this purpose, one may split the Hamiltonian as

$$H = H_0' + V', \quad H_0' \equiv -\Delta/2 + q^2/2, \quad V' \equiv V - q^2/2$$

and one may write formulas similar to those discussed above, with Φ_0 the ground state of the new "free" Hamiltonian H_0' and $dW_{x,x'}$ replaced by the conditional functional measure associated with the Ornstein-Uhlenbeck velocity process, [27] i.e. with the positive kernel of $e^{-\tau H_0'}$.

The above equation for the Schwinger functions is the imaginary time (or euclidean) version of the Gell-Mann-Low formula [28] and expresses the solution of the quantum mechanical problem as a quadrature, in terms of a functional integral.

In the theoretical physics literature the path integral formula for the Schwinger functions is often written in the (heuristic) Feynman path-integral form

$$< x(\tau_1), ..., x(\tau_n) > = Z^{-1} \int \mathcal{D}x(\cdot)\, e^{-\int ds\, [m\,\dot{x}(s)/2 + V(x(s))]}\, x(\tau_1), ..., x(\tau_n),$$

$$Z \equiv \int \mathcal{D}x(\cdot)\, e^{-S^E(x(\cdot))}, \quad S^E(x(\cdot)) = \int ds\, [m\dot{x}(s)/2 + V(s)], \qquad (6.5.8)$$

where $S^E = \mathcal{T} + \mathcal{V}$ is the imaginary time (or euclidean) action; instead of the Wiener measure the formula involves the ill defined product of $\mathcal{D}x(\cdot)$ and $e^{-\mathcal{T}}$, and the non L^2 function $\Phi_0(x) = 1$ is used as boundary condition.

6.6   Euclidean quantum mechanics

As we have seen in the previous section, the ground state correlation functions at imaginary times provide complete information on the quantum mechanical system and have an interpretation in terms of a stochastic process. This approach to quantum mechanics, which exploits the wisdom of the theory of stochastic processes and expresses the solution as a functional integral has received much attention over the last years. The functional integral approach has proved very useful for the discussion of important quantum mechanical problems as well as for non-perturbative insights about the solutions, also in the infinite dimensional case of quantum field theory or many body theory. [29]

The recognition of the power of such an approach to quantum mechanics is probably at the roots of the point of view that such an alternative formulation of quantum mechanics supersedes the Heisenberg operator approach. Actually, it has even become fashionable to formulate quantum mechanical models directly in the imaginary time or euclidean version, without a careful check of their being well posed as real time quantum mechanical

[27] For a more detailed discussion see J. Glimm and A. Jaffe, *Quantum Physics. A Functional Integral Point of View*, Springer 1987, Sects. 3.2-3.4.

[28] M. Gell-Mann and F. Low, Phys. Rev. **84**, 350 (1951).

[29] See B. Simon, *Functional Integration and Quantum Mechanics*, Academic Press 1979; S. Albeverio, in *Proceedings of Symposia in Applied Mathematics Vol.52*, Am. Math. Soc. 1997. A brief discussion of the usefulness of the functional integral approach is given in the next section.

models. This problem is not merely academic, because in the manipulations or approximations of the functional integral one should keep under control the fulfillment of the general properties which guarantee the recovery of a real time interpretation. This section is devoted to the discussion of the general properties of the Schwinger functions which guarantee that they arise by analytic continuations to imaginary times of quantum mechanical models.

As a trivial consequence of the analogous property of the Wightman functions we have

S1. (*Translation invariance*)

$$S_\alpha(\tau_1 + s, ..., \tau_n + s) = S_\alpha(\tau_1, ..., \tau_n). \qquad (6.6.1)$$

This means that the corresponding stochastic process is stationary.

By the functional integral representation of the Schwinger functions one gets their symmetry under permutations π of their arguments

S2. (*Symmetry*)

$$S_{\alpha_{\pi(1)}...\alpha_{\pi(n)}}(\tau_{\pi(1)}, ..., \tau_{\pi(n)}) = S_{\alpha_1...\alpha_n}(\tau_1, ..., \tau_n). \qquad (6.6.2)$$

From a conceptual point of view such a property has far reaching consequences, because it implies that euclidean quantum mechanics is described by an *abelian* algebra of (random) variables. This has indeed been emphasized as the revolutionary feature of the Feynman path integral approach, especially for his strong relation with classical (statistical) mechanics. It should be stressed that such an abelian structure has been obtained for the imaginary time version and that the non-abelian structure of quantum mechanics is here encoded in the different boundary values taken by the Schwinger functions when one approaches the real axis from above and from below (see e.g. the case of the two point function).

By definition, the Schwinger functions are analytic functions of their arguments for chronologically ordered times and so are also their symmetric extensions to unordered (imaginary) times, if the times are non-coincident, i.e. $\tau_j \neq \tau_i$ if $j \neq i$, or if $s_j = \tau_{j+1} - \tau_j \neq 0$, $\forall j$. Furthermore, for chronologically ordered times the Schwinger functions are the Laplace transforms of the Wightman functions, which are tempered distributions. This implies that the Schwinger functions have at most polynomial singularities at coincident times. The mathematical characterization of such properties yields the following condition. [30]

---

[30] See B. Simon, *The $P(\phi)_2$ Euclidean (Quantum) Field Theory*, Princeton Univ. Press 1974, Theor. II.3, Theor. II.5, Theor. II.8; for a handy account see e.g. F. Strocchi, *Selected Topics On The General Properties Of Quantum Field Theory*, World Scientific 1993, Sect. 3.5.

S3. *(Laplace transform condition)*

$$S_\alpha(s_1, ..., s_{n-1}) \in \mathcal{S}(\mathbf{R}_+^{n-1})' \tag{6.6.3}$$

where $\mathcal{S}(\mathbf{R}_+^{n-1})'$ denotes the continuous linear functionals on the family $\mathcal{S}(\mathbf{R}_+^{n-1})$ of test functions in $\mathcal{S}(\mathbf{R}^{n-1})$ with support in $\mathbf{R}_+^{n-1}$. [31]

The next important property is the euclidean counterpart of the positivity condition of the Wightman functions. For this purpose, putting

$$\Psi_n(\tau, s_1, ..., s_n) \equiv e^{-\tau H} e^{i\alpha_1 q} e^{-s_1 H} ... e^{-s_{n-1}H} e^{i\alpha_n q} \Psi_0 \equiv \Psi_n(\tau, s),$$

we have

$$(\Psi_n(\tau, s), \Psi_m(\tau', s')) = S(s_{n-1}, ..., s_1, \tau + \tau', s'_1, ..., s'_{m-1})$$

$$= \mathcal{S}(-\tau_n, ..., -\tau_1, -\tau, \tau', ..., \tau'_m).$$

Therefore, by the positivity of the Hilbert scalar product, one has

S4. *(Reflexion Positivity)* For any terminating sequence $\{f_j \in \mathcal{S}(\mathbf{R}_+^j)\}$, putting $\theta f_j(\tau_1, ..., \tau_j) \equiv f_j(-\tau_1, ..., -\tau_j)$,

$$\sum_{j,k=0}^{N} \int d\tau_1 ... d\tau_{j+k} \overline{\theta f}_j(\tau_j, ..., \tau_1) S_{\alpha_{jk}}(\tau_1, ..., \tau_j; \tau_{j+1}, ..., \tau_k) f_k(\tau_{j+1}, ..., \tau_k) \geq 0,$$

$$S_{\alpha_{jk}}(\tau_1, ..., \tau_j; \tau_{j+1}, ..., \tau_k) \equiv S_{-\alpha_1...-\alpha_j, \alpha_{j+1}...\alpha_k}(\tau_1, ..., \tau_j, \tau_{j+1}, ..., \tau_k).$$

This crucial property was discovered by Osterwalder and Schrader and for this reason it is also called the *OS-positivity condition*. It is thanks to this property that one recovers a Hilbert space of states from the Schwinger functions and an acceptable real time interpretation. [32]

A family of Schwinger functions satisfying **S1-S4** define what can be called *euclidean quantum mechanics*. The proof that **S1-S4** do guarantee the reconstruction of the real time quantum mechanics, may be done along the same lines of the real time (Wightman) reconstruction theorem, by a GNS type argument. [33]

For the class of quantum mechanical models discussed above, for which the kernel of $e^{-\tau H}$ is strictly positive (see Sect. 4), the Schwinger functions

---

[31] We do not dwell on this technical (but relevant) point, for which we refer to the references of the preceding footnote.

[32] For a characterization of the stochastic processes associated with the euclidean version of quantum mechanics, in particular for the relation between the OS-positivity condition and the Markov property, and for the discussion of the associated semigroup structure see A. Klein, Jour. Funct. Anal. **27**, 277 (1978) and B. Simon, *The $P(\phi)_2$ Euclidean (Quantum) Field Theory*, Princeton Univ. Press 1974.

[33] For the details of the proof see e.g. the references listed in footnote 30.

satisfy the following additional property

S5. (*Nelson positivity*) The Schwinger functions

$$S_\alpha(s_1, ..., s_{n-1}) = (\Psi_0, e^{i\alpha_1 q} e^{-s_1 H} ... e^{-s_{n-1} H} e^{i\alpha_n q} \Psi_0)$$

define a positive linear functional ω_E on the euclidean algebra $\mathcal{A}_E$ generated by the functions $U(\alpha, \tau) = e^{i\alpha x(\tau)}$, (closed in the sup norm).

This follows from the positivity of the ground state wave function $\Psi_0(x)$ and the strict positivity of the kernel of $e^{-\tau H}$ (see Sect. 6.4). This positivity property is also easily readable from the functional integral representation of ω_E

$$\omega_E(F(x(\tau_1), ..., x(\tau_n)))$$

$$= \int dx\, dx'\, \overline{\Psi_0}(x) \Psi_0(x') \int dW_{x,x'}(x(\cdot))\, e^{-\int_{-T}^{T} V(x(s))ds}\, F(x(\tau_1), ..., x(\tau_n)),$$

which is clearly positive if so is F.

Quite generally, the property of Nelson positivity by itself guarantees that the Schwinger functions can be written in terms of a (positive) functional measure over the space of trajectories as discussed in Appendix J.

Whereas OS-positivity is crucial for the real time interpretation and therefore it cannot be dispensed with, Nelson positivity is less compulsory and in fact may fail in some quantum mechanical models, for which a possible functional integral representation may require complex measures.[34]

---

[34]See e.g. J. Löffelholz, G. Morchio and F. Strocchi, Ann. Phys. **250**, 367 (1996).

6.7 Applications of the functional integral

The conceptual aspects of the functional integral approach to quantum mechanics have been discussed in the previous sections. Now we would like to argue in favor of its usefulness in solving quantum mechanical problems and to point out its advantages with respect to the operator approach, especially in providing non-perturbative information.

6.7.1 Feynman perturbative expansion

Historically, one of the main motivations for the functional integral approach to quantum mechanics is that it yields a simple and compact perturbative expansion (*Feynman perturbative expansion*).

The idea is to expand the exponential $\exp - \int_{-T}^{T} V(x(s)) \, ds$ in the functional integral representation of the ground state correlation functions (or of the kernel $\mathcal{K}$ of $e^{-\tau H}$) and to reduce the computation of each order to Gaussian integration. In fact, by performing such an expansion in the analog of eq. (6.5.7), one has a functional integral of the form

$$\int d\mu_W(x(\cdot)) \sum_k (-1)^k (k!)^{-1} \left(\int_{-T}^{T} V(x(s)) \, ds \right)^k x(\tau_1)...x(\tau_n),$$

and by an interchange of integrations (Fubini's theorem) one gets terms of the form

$$\int_{-T}^{T} ds_1...ds_k \int d\mu_W(x(\cdot)) \, V(x(s_1))...V(x(s_k)) x(\tau_1)...x(\tau_n).$$

By exploiting the symmetry of the integrand in the variables $s_1,...s_k$ one may reduce the integration to the ordered sector $s_1 \leq s_2 \leq ... \leq s_k$.

$$k! \int_{-T}^{T} ds_1 \int_{s_1}^{T} ds_2... \int_{s_{k-1}}^{T} ds_k \int d\mu_W \, V(x(s_1))...V(x(s_k)) \, x(\tau_1)...x(\tau_n).$$

By the symmetry of the Schwinger functions one may consider the case of ordered times $\tau_1 \leq ... \leq \tau_n$, split the original integration interval $(-T,T)$ into $n+1$ subintervals $(-T,\tau_1),...,(\tau_n,T)$ and get a series each term of which involves a Gaussian integration of a time ordered product of random variables. When V is a polynomial, such a Gaussian integration is easily done with the help of Wick's theorem (see Appendix I). This is Feynman (euclidean) perturbative expansion. [35]

---

[35] R.P. Feynman, Rev. Mod. Phys. **20**, 367 (1948). For a mathematical discussion of the perturbative expansion of the euclidean functional integral see B. Simon, *Functional Integral and Quantum Physics*, Academic Press 1979, Sect. 20.

The same expansion can be applied to the functional integral representation of the kernel $\mathcal{K}(x, \tau; x', \tau')$, eq. (6.4.1), and for the k-th order term one gets

$$(-1)^k \int_{\tau'}^{\tau} ds_1 ... \int_{s_{k-1}}^{\tau} ds_k \int dW_{x,x'} V(x(s_1))...V(x(s_k)),$$

where $dW_{x,x'}$ denotes the conditional Wiener measure obtained by fixing the path end points $x(\tau) = x, x(\tau') = x'$. By an easy extension of the Feynman-Kac formula, essentially given by the functional integral representation of the kernel of

$$e^{-(\tau-s_1)H_0} A_1(q) e^{-(s_1-s_2)H_0} ... A_k(q) e^{-(s_k-\tau')H_0}$$

(with $A_j(q)$ bounded function of q), as discussed in Sect.6.5, one may write the k-th order term as

$$(-1)^k \int_{\tau'}^{\tau} ds_1 ... \int_{s_{k-1}}^{\tau} ds_k \int dx_1 ... dx_k K(x, t; x_1, s_1)$$

$$V(x_1) K(x_1, s_1; x_2, s_2) V(x_2)...K(x_k, s_k; x', \tau'). \tag{6.7.1}$$

Equation (6.7.1) has a suggestive pictorial representation obtained by drawing a segment in space time with end points $(x, \tau), (y, s)$ for each propagator $K(x, \tau; y, s)$ and a bubble at (y, s) for the factor $V(y)$. The so obtained diagram is a *Feynman diagram* pictorially "representing" a Brownian propagation from (x, t) to (x_1, s_1), where an interaction occurs with $V(x_1)$, followed by a propagation from (x_1, s_1) to (x_2, s_2), etc. The visualization of the perturbative expansion in terms of Feynman diagrams provides an easy bookkeeping of the various terms.

The same strategy can be applied to the Feynman path-integral for the real time propagator and for the k-th term one gets the same formula as above, with the imaginary time kernel K replaced by the real time kernel G. In this way one gets the real time perturbative expansion, as given by Feynman, as well as its pictorial representation in terms of (real time) Feynman diagrams. They have a pictorial interpretation in terms of free propagation of the particle from, say, x to y_1 in the time interval $s_1 - t$, where it interacts with the potential $V(y_1, s_1)$ and then propagates freely to y_2, where it interacts with the potential $V(y_2, s_2)$ etc., the n-th order involving n interactions. The actual propagation is then the result of summation over all such motions, with the interaction points integrated over. [36] Each order

---

[36]R.P. Feynman, Phys. Rev. **76**, 749 (1949); Phys. Rev. **76**, 769 (1949); R.P. Feynman and A.R. Hibbs, *Quantum Mechanics and Path Integrals*, McGraw-Hill 1965, Chap. 6 . For an elementary account of the case of non-relativistic quantum mechanics see e.g. E. Corinaldesi and F. Strocchi, *Relativistic Wave Mechanics*, North-Holland 1963, Chap. IV, and L. Schulman, *Techniques and Applications of Path Integration*, Wiley 1981, Chap. 10.

of such a perturbative expansion, eq. (6.7.1), does no longer involves path integration, and it is mathematically well defined without the problems of the Feynman path integral. This fact has led to the point of view, sometimes adopted in the literature, of defining the Feynman path integral by its perturbative expansion. The drawbacks of this position are the lack of control of the convergence of the perturbative series, which generically does not converges and in the most favorable cases has to be understood as an asymptotic series, and the fact that in this way one looses one of the main virtues of the functional integral approach, namely its non-perturbative character.

6.7.2   Semiclassical limit

Another virtue of the functional integral approach is to provide a simple discussion of the classical limit and furthermore it allows to take advantage of the classical solutions for the quantum mechanical problem. For example, one may develop a perturbative expansion in which non-linear and/or non-perturbative effects encoded in the classical solutions can be taken into account at zero order of the expansion.

The classical limit is defined by $\hbar \to 0$ and it can be discussed at the level of the functional integral by the infinite dimensional version of Laplace's method or saddle-point approximation. [37]

The idea is that, in the limit $\hbar \to 0$, an integral of the form

$$I(\hbar) = \int_a^b dx\, f(x) e^{-s(x)/\hbar},$$

with $s(x) \geq 0$, $f(x)$ and $s(x)$ regular in $[a, b]$, is dominated by the minima of $s(x)$ and by the quadratic expansion of $s(x)$ around its minima . This can be easily seen in the one dimensional case, since, if the interval $[a, b]$ does not contain stationary points of $s(x)$, i.e. $s'(x) \neq 0$ there, then one can make a (invertible) change of variables $z = s(x)/\hbar$, $x = x(z)$, and obtain

$$\hbar \int_{s(a)/\hbar}^{s(b)/\hbar} e^{-z} f(x(z))/s'(x(z))\, dz \approx_{\hbar \to 0} \hbar\, e^{-s(\bar{x})/\hbar}\, f(\bar{x})/s'(\bar{x}),$$

where by the mean value theorem $\bar{x}$ is a point of $[a, b]$. On the other hand, if x_0 is a minimum of $s(x)$, by restricting the integral to $[x_0 - \epsilon, x_0 + \epsilon]$ and by expanding $s(x)$ around x_0 and by a change of variables $y = \hbar^{-1/2}(x - x_0)$ one obtains

$$\sqrt{\hbar} \int_{-\epsilon/\sqrt{\hbar}}^{\epsilon/\sqrt{\hbar}} [f(x_0) + O(\sqrt{\hbar})] e^{-[s(x_0)/\hbar + s''(x_0)y^2/2 + O(\sqrt{\hbar})]}\, dy,$$

---

[37]See A. Erdélyi, *Asymptotic Expansions*, Dover 1956, Sect. 2.4; F.W.J. Olver, *Introduction to Asymptotics and Special Functions*, Academic Press 1974, Sect. 3.7.

which for $\hbar \to 0$ (by a Gaussian integration) behaves like

$$\sqrt{\hbar}\, f(x_0) e^{-s(x_0)/\hbar} (2\pi/s''(x_0))^{1/2}$$

and dominates over the previous term (as $s(\bar{x}) - s(x_0) > 0$). By the same reason stationary points, which are not minima, give subdominant contributions to the asymptotic behaviour.

The same argument can be used for the n-dimensional case; the quadratic term in the expansion of $s(x)$ gives now a quadratic form $(y, \partial^2 S y)$ with $\partial^2 S$ the matrix of the second derivatives of s at the minimum. Again the behaviour for $\hbar \to 0$ is obtained by a Gaussian integration and one has the following asymptotic behaviour

$$\sqrt{\hbar}\, e^{-s(x_0)/\hbar} (2\pi)^{n/2} (\det \partial^2 S)^{-1/2}.$$

The functional integral case can be heuristically obtained by using the formal expression (6.5.8) and by expanding the euclidean action around the minimizing trajectories $x_{cl}(\tau)$, i.e. the solutions of the classical *euclidean* equations, with appropriate boundary conditions.

In this way, one gets the following asymptotic behaviour, for $\hbar \to 0$, for the kernel of $e^{-Ht/\hbar}$

$$\mathcal{K}(x, x'; \tau, 0) \sim N\sqrt{\hbar}\, e^{-S_{cl}^E(x, x', \tau)/\hbar} (\det \partial^2 S_{cl}^E(\tau))^{-1/2},$$

where N is a normalization constant, which can be fixed by comparison with the explicit expression of the free kernel

$$K(x, x; \tau, 0) = (m/2\pi\tau\hbar)^{1/2} \sim N\sqrt{\hbar}\, (\det \partial^2 S_{0,\, cl}^E(\tau))^{-1/2}.$$

In conclusion, one has

$$\mathcal{K}(x, x'; \tau, 0) \sim \left(\frac{m}{2\pi\tau\hbar}\right)^{1/2} e^{-S_{cl}^E(x, x', \tau)/\hbar} \left(\frac{\det \partial^2 S_{0,cl}^E(\tau)}{\det \partial^2 S_{cl}^E(\tau)}\right)^{1/2}. \qquad (6.7.2)$$

A more convincing argument can be obtained by working on the polygonal path approximation and by expanding the n-th order finite difference action S_n^E around its minima, which are given by the solutions of the finite difference (euclidean) equations of motion. As above, one may argue that the quadratic term of the expansion gives the leading contribution with respect to the higher order terms. Thus, one ends up with a Gaussian integration leading to the determinant of the finite dimensional second order variation of S_n^E, as in the n-th dimensional case discussed above. Thus, for the n-th order expression of the propagator one gets

$$K^{(n)}(x, x'; \tau, 0) \sim (m/2\pi\hbar\tau)^{1/2} e^{-S_n^E/\hbar} (\det S_{0,n}^E(\tau)/\det S_n^E(\tau))^{1/2}.$$

The control of the limit $n \to \infty$ is made easier by the fact that the ratio of the two determinants is better behaved than each of them (this can be regarded as an example of *renormalization*). The limit of the determinant ratio $r(\tau)$ can be computed explicitly [38] and it is given by $r(\tau) = \tau/\Delta(\tau)$ with $\Delta(\tau)$ solution of the equation

$$m\, d^2 \Delta(\tau)/d\tau^2 + V''(x_{cl}(\tau))\, \Delta(\tau) = 0,$$

with boundary conditions $\Delta(0) = 0$, $\Delta'(0) = 1$.

The above formula (6.7.2) for the semiclassical limit displays a deep connection between quantum mechanics and classical mechanics; in particular it provides a strong support and clarification of Dirac proposal. The saddle point approximation discussed above yields the exact result when the action is at most quadratic. It is very instructive to compute the kernel $\mathcal{K}$ in the case of the harmonic oscillator. The result is

$$\mathcal{K}(x, x'; \tau, 0) = \left(\frac{m\omega}{2\pi\hbar \sinh(\omega\tau)} \right)^{1/2} e^{-S_{cl}(x,x';\tau)},$$

$$S_{cl} = (m\omega/2 \sinh(\omega\tau)) \left[(x'^2 + x^2) \cosh(\omega\tau) - 2x\, x' \right].$$

To deduce the above formula one remarks that the differential operators $\Delta \equiv -d^2/ds^2 + \omega^2$, and $\Delta_0 \equiv -d^2/ds^2$, $s \in [0, \tau]$, with boundary conditions $\psi(0) = \psi(\tau) = 0$, have the following eigenvalues $(n\pi/\tau)^2 + \omega^2$ and $(n\pi/\tau)^2$, $n = 1, 2, ...$, respectively, (corresponding to the eigenfunctions $\psi_n(s) = (2/\tau)^{1/2} \sin(n\pi s/\tau)$) and therefore

$$\det \Delta / \det \Delta_0 = \sum_{n>0} (1 + (\omega\tau/n\pi)^2) = \sinh \omega\tau/(\omega\tau).$$

The semiclassical limit shows that even if the classical solutions, which are continuously differentiable, belong to a set of zero measure, nevertheless they dominate the classical limit. This indicates that the irregular behaviour of the relevant trajectories is a "quantum" effect and disappears in the limit $\hbar \to 0$; it can be viewed as due to the rapid fluctuations around the classical trajectory. From a mathematical point of view, one is facing the intriguing situation in which a set of points of zero measure determine the asymptotic ($\hbar \to 0$) behaviour of the functional integral.

---

[38] I.M. Gelfand and A.M. Yaglom, Jour. Math. Phys. **1**, 48 (1960); J.H. Van Vleck, Proc. Nat. Acad. Sci. **14**, 178 (1928); W. Pauli, *Pauli Lectures on Physics, Selected Topics in Field Quantization*, C.P. Enz ed., MIT Press 1973; S. Coleman, *Aspects of Symmetry*, Cambridge Univ. Press 1985, Chap. 7; L.S. Schulman, *Techniques and Applications of Path Integration*, Wiley 1981. For a rigorous discussion of the semiclassical limit of the euclidean functional integral see B. Simon, *Functional Integration and Quantum Physics*, Academic Press 1970, Chap. VI.

Another interesting feature of the semiclassical limit is that it provides an alternative perturbative expansion in powers of $\hbar$, rather than in powers of the strength of the potential. The lowest order is given by the quadratic term in the expansion of the action around its minimum and it is not necessarily given by the theory with $V = 0$ (or with the quadratic term of the potential). In this way, one builds up an expansion in which non-perturbative features, encoded in the classical solutions, can be taken into account at zero order.

For example, for the anharmonic oscillator defined by the potential $V = \mu^2 x^2 + \lambda^2 x^4, \lambda \geq 0$, the above expansion at the lowest order leads to the Orstein-Uhlenbeck process, rather than to the Wiener process. Even more interesting is the double well potential $V = \lambda(x^2 - a^2)^2$, for which the lowest order of the above expansion is given by the quadratic expansions around the two absolute minima $\bar{x} = \pm a$, rather than the quadratic expansion around $\bar{x} = 0$. In quantum field theory and in many body theory, such an expansion corresponds to the so-called loop expansion [39].

The semiclassical limit can be discussed also for the Feynman path integral. In this case the limit $\hbar \to 0$ is governed by the stationary phase approximation, by which

$$\int_a^b f(x) e^{is(x)/\hbar}\, dx \sim \sqrt{\hbar}\, f(x_0) e^{is(x_0)/\hbar} \left(2i\pi/s''(x_0)\right)^{1/2},$$

where x_0 is a stationary point of $s(x)$ in $[a, b]$. [40] The infinite dimensional case is however more delicate and more difficult than for the saddle point approximation. Heuristic arguments can be found in the book by Schulman quoted before; a rigorous approach has been undertaken by Albeverio and collaborators. [41] The nice feature of the Feynman path integral is that the semiclassical limit is dominated by the solution of the real time classical equations of motion.

6.7.3 Ground state properties

By using the spectral theorem, one can easily show that

$$\text{strong} - \lim_{t \to \infty} e^{-t(H - E_0)} = P_0,$$

---

[39] See S. Coleman, *Aspects of Symmetry*, Cambridge Univ. Press 1985, Chap. 5, Sect. 3.4, where one can also find a clear discussion of the merits of such an expansion. For the advantages of the loop expansion, with respect to the ordinary perturbative expansion in the strength of the potential, in many body theory see J.W. Negele and H. Orland, *Quantum Many-Particle Systems*, Addison-Wesley 1988, esp. Sect. 2.5.

[40] If $s'(x) \neq 0$ in [a,b], by a change of variable $z = s(x)$ and by partial integration one sees that the integral decreases like $\hbar$. For an integral around the stationary point x_0, one expands $s(x)$ and performs a Gaussian-like integration as before.

[41] S. Albeverio and Z. Brzeźniak, Ann. Phys. **113**, 177 (1993) and references therein.

where E_0 is the ground state energy and P_0 the projection on the ground state.

Thus, for any state ψ with a non-trivial spectral support on $[E_0, E_0 + \delta]$, $\delta > 0$, one has

$$(\psi, e^{-tH} \psi) \approx_{t \to \infty} e^{-tE_0} |(\psi, \Psi_0)|^2$$

and therefore

$$E_0 = - \lim_{t \to \infty} t^{-1} \log(\psi, e^{-tH} \psi),$$

$$|(\psi, \Psi_0)|^2 = \lim_{t \to \infty} e^{tE_0}(\psi, e^{-tH} \psi).$$

Since the right hand sides of the above equations have a functional integral representation, one can in this way get information on the ground state energy and wave function. For this purpose, one can e.g. use the semiclassical expansion discussed above. [42]

6.7.4   Coupling constant analyticity

As already remarked, the functional integral is very powerful for non-perturbative questions, since, in contrast with the perturbative expansion, it displays the solution as an integral.

A typical question is whether the perturbative expansion converges and/or what type of information about the exact solution can be derived from it. To illustrate this problem, we consider the anharmonic oscillator defined by the potential $V = \mu x^2 + g x^4$, $\mu, g \in \mathbf{R}_+$, (which is also the simplest prototype of scalar field theory and corresponds to the ϕ^4 model in 0+1 dimensions). The question is the status of the perturbative expansion in g around the point $g = 0$, of the euclidean correlation functions. One expects that the so obtained series does not converge, since for $g < 0$ the Hamiltonian is unbounded from below, instability occurs and this does not fit with analyticity in g at the origin. [43] One can actually show that the coefficients of the power series increase faster than $n!$ and that indeed the perturbative series diverges. The series is however asymptotic to the solution and therefore it provides non trivial information. [44]

To give the flavor of the effectiveness of the functional integral for the above problems we briefly discuss the analyticity properties of an ordinary

---

[42]Simple interesting examples of this strategy are discussed in S. Coleman, *Aspects of Symmetry*, Cambridge Univ. Press 1985, Chap. 7, Sects. 2-2.4. General quantum mechanical problems are discussed in B. Simon, *Functional Integration and Quantum Physics*, Academic Press 1979.

[43]A similar argument about the necessary divergence of the pertubative series in Quantum Electrodynamics has been discussed by F. Dyson, Phys. Rev. **85**, 631 (1952).

[44]See B. Simon, *Functional Integration and Quantum Physics*, Academic Press 1979, Sect. 20; A.S. Wightman, in *Mathematical Quantum Field Theory and Related Topics*, Montreal 1977, J.S. Feldman and L.M. Rosen eds., Am. Math. Soc. 1988, p. 1.

integral of the form

$$Z(\mu, g) \equiv \int dx e^{-\mu x^2 - g x^4}, \quad \mu > 0,$$

which mimics the partition function of a zero dimensional ϕ^4 model.

The perturbative series in powers of g reads

$$\sum_n g^n Z_n(\mu), \quad Z_n(\mu) = (n!)^{-1}(-1)^n \mu^{-2n-1/2} \int e^{-y^2}(y^2)^{2n-1/2} dy^2/2.$$

The last integral is related to the Γ function

$$\Gamma(2n + 1/2) = 2\sqrt{\pi} \, 2^{-4n}(4n)!/(2n)!$$

which increases like $(2n)!$ for large n. Hence, the coefficients of the series increase like $2^n n!$ and the series cannot converge, in agreement with Dyson argument. By estimating the rest of the Taylor series, however, we get

$$\left| Z(\mu, g) - \sum_{n=0}^{k} g^n Z_n \right| \leq C_k g^{k+1} |Z_{k+1}|,$$

i.e. the perturbative series is asymptotic to the exact solution. [45]

For $\mu < 0$ the model mimics the double well potential; to discuss its analyticity properties it is convenient to complete the square and consider the modified "action" $g(x^2 + \mu/2g)^2$, which differs from the original one by the factor $\exp(-\mu^2/4g)$. With this modification of the action one gets a perturbative series in powers of $4g/\mu^2$ which is asymptotic to the solution, but the original one is not. [46]

---

[45] We recall that a series $\sum a_n x^n$ is asymptotic at the origin to a function $f(x)$, which is C^∞ in $(0, \epsilon)$, if $\forall k \; f(x) - \sum_{n=0}^{k} a_n x^n = o(x^k)$ for $x \to 0^+$. This implies that $a_n = (n!)^{-1} d^n f(0^+)/dx^n$.

[46] For a detailed analysis of this very simple, but nevertheless instructive model, we refer to Wightman contribution quoted above.

6.8   Functional integral and topology

The functional integral approach is particularly useful for the quantization of systems with constraints, since their fulfillment can be realized by suitably restricting the functional measure to the admissible trajectories, [47] whereas the Hamiltonian approach is in general less direct and more cumbersome.

The functional integral also proves useful for discussing quantum mechanics on manifolds, e.g. a quantum particle living on a (closed) manifold. The problem is not trivial because the local translations on the manifold do not automatically exponentiate to give the Weyl relations and topological constraints may intervene. To show the effectiveness of the functional integral approach we shall discuss the example of a **quantum particle on a circle**. The example may appear physically trivial, but it is instead very instructive for the general strategy and for the new phenomena which appear: i) the occurrence of *superselection rules*, namely the existence of a non trivial center of the observables, ii) the existence of *inequivalent representations of the observable algebra*, in contrast with Von Neumann uniqueness theorem of ordinary quantum mechanics, iii) the existence of automorphisms of the observable algebra which are not implemented by unitary operators in irreducible representations of the observable algebra (the so-called phenomenon of *spontaneous symmetry breaking*) iv) the interplay between topology and quantization. The model is also a prototype of phenomena which play a crucial role in gauge quantum field theory.

1. The C*-algebra. The canonical variables, which describe a classical particle of mass m on a circle of radius R, are the angle ϕ, $0 \leq \phi < 2\pi$, and the conjugated momentum $p_\phi = mR^2\dot{\phi}$, hereafter denoted by p. Since ϕ is defined only modulo 2π only periodic functions of ϕ may describe observable quantities and the role of the Weyl algebra of ordinary quantum mechanics is now taken by the C^*-algebra $\mathcal{A}$ generated by $e^{in\phi}$, $e^{i\beta p}$, $n \in \mathbf{Z}, \beta \in \mathbf{R}$. The algebraic structure is given by the Weyl relations (eqs. (3.1.3), (3.1.4)) with α replaced by n).

The algebra $\mathcal{A}$ has a non-trivial center $\mathcal{Z}$ generated by $V(2\pi) \equiv e^{i2\pi p}$ and therefore $\mathcal{A}$ does not have irreducible faithful representations. Each irreducible representation π is labeled by an angle $\theta \in [0, 2\pi)$, (see below), defined by $\pi_\theta(e^{i2\pi p}) = e^{i\theta}$. Von Neumann uniqueness theorem does not apply and in fact one has inequivalent irreducible representations of $\mathcal{A}$, strongly continuous in β, briefly called *regular*.

2. The representations. In order to classify the irreducible regular representations of $\mathcal{A}$, we note that the representations π_θ, with $\pi_\theta(e^{i2\pi p}) =$

---

[47]L.D. Faddeev, Theoret. Math. Phys. **1**, 1 (1970); P. Senjanovic, Ann. Phys. **100**, 227 (1976); C. Teitelboim and M. Henneaux, *Quantization of Gauge Systems*, Princeton Univ. Press 1992.

$e^{i\theta}$, are related by the automorphisms ρ^θ

$$\rho^\theta(W(n,\beta)) \equiv e^{i\tilde\theta\,\beta}\,W(n,\beta), \quad \tilde\theta \equiv \theta/2\pi \qquad (6.8.1)$$

corresponding to $W(\tilde\theta,0)\,W(n,\beta)\,W(-\tilde\theta,0)$. In fact, for any π_θ, the representation

$$\pi_0(\mathcal{A}) \equiv \pi_\theta(\rho^{-\theta}(\mathcal{A}))$$

satisfies

$$\pi_0(e^{i2\pi p}) = \mathbf{1}\,.$$

If we show that the irreducible representations π of $\mathcal{A}$, with $\pi(e^{i2\pi p}) = \mathbf{1}$ are all unitarily equivalent, we have also shown that any π_θ is uniquely determined by θ (up to isomorphisms), since it is given by

$$\pi_\theta(A) = \pi_0(\rho^\theta(A)), \quad \forall A \in \mathcal{A}.$$

For this purpose, as in the proof of Von Neumann theorem, we show that for any irreducible regular representation π with $\pi(e^{i2\pi p}) = \mathbf{1}$, the corresponding Hilbert space $\mathcal{H}_\pi$ contains a (cyclic) vector Ψ_0 such that

$$(\Psi_0,\, \pi(W(n,\beta))\,\Psi_0) = \delta_{n,0}.$$

Now, for any such representation $\pi(W(0,\,\beta)) = \pi(W(0,\,\beta+2\pi))$, i.e. $\pi(W(0,\,\beta))$ is a periodic function. Thus, the operator

$$P \equiv (2\pi)^{-1} \int_0^{2\pi} d\beta\,\pi(W(0,\beta)) = P^*$$

is well defined and it cannot vanish because

$$\pi(W(-n,-\gamma))\,P\,\pi(W(n,\gamma)) = (2\pi)^{-1} \int d\beta\,\pi(W(0,\,\beta))\,e^{i\,n\,\beta}$$

and the vanishing of the right hand side would imply the vanishing of the periodic function $\pi(W(0,\,\beta))$, which is excluded by the unitarity of $\pi(W(0,\,\beta))$.

Furthermore,

$$P\,\pi(W(n,\beta))\,P = (2\pi)^{-2} \int_{-\pi}^{\pi} d\gamma\,e^{in\gamma} \int_0^{2\pi} d\gamma'\,\pi(W(0,\,\beta+2\gamma')) = \delta_{n,0}.$$

The proof then continues as in the case of Von Neumann theorem.

The representation π_0 defines an abstract C^*-algebra $\mathcal{A}_{per}$ isomorphic to $\pi_{\theta=0}(\mathcal{A})$, which may be identified with the C^*-algebra generated by $W(n,0)$ and $W_{per}(0,\beta)$ with $W_{per}(0,\beta+2\pi) = W_{per}(0,\beta)$,

$$W_{per}(0,\beta)\,W_{per}(0,\beta') = W_{per}(0,\beta+\beta') = W_{per}(0,\beta+\beta'\,\mathrm{mod}\,2\pi),$$

$$W_{per}(0, \beta)\, W(n, 0) = W(n, 0)\, W_{per}(0, \beta)\, e^{i\beta n} \equiv e^{i\beta n/2}\, W_{per}(n, \beta).$$

The above generalization of Von Neumann theorem shows that all the irreducible regular representations of $\mathcal{A}_{per}$ are unitarily equivalent.

The analog of the Schroedinger representation of $\mathcal{A}_{per}$ is given by the representation space $L^2([0, 2\pi), d\phi)$, with $\pi(W(n, 0))$ acting as the multiplication operator $e^{in\phi}$ and $\pi(W_{per}(0, \beta))$ acting as a rotation $\phi \to \phi + \beta$ mod 2π, with generator $p = -i\partial/\partial\phi$ on periodic functions (see Appendix F). This defines the Schroedinger representation of $\mathcal{A}$ with $\theta = 0$.

The corresponding representations of $\mathcal{A}$ with $\theta \neq 0$ are then given by

$$(\pi_\theta(W(0, \beta))\,\psi)(\phi) = (\pi_0(\rho^\theta(W(0, \beta))\,\psi)(\phi)$$

$$= e^{i\beta\tilde{\theta}}\,(\pi_0(W(0, \beta))\psi)(\phi) = \psi(\phi + \beta \bmod 2\pi)\, e^{i\beta\,\tilde{\theta}}. \qquad (6.8.2)$$

Representations with different θ correspond to different self-adjoint extensions, denoted by p_θ, of the differential operator $-i\partial/\partial\phi$, and are labeled by the angle appearing in the boundary conditions $\psi(2\pi) = \psi(0)e^{i\theta}$, (see Appendix F). It is clear from the above equations that the above Schroedinger representation with $\theta = 0$ corresponds to $p = -i\partial/\partial\phi$ on periodic functions, whereas the corresponding representation with $\theta \neq 0$ corresponds to a realization of the generator as $p_\theta = -i\partial/\partial\phi + \theta/2\pi$ on periodic functions. A unitarily equivalent representation is given by ($[\,.\,]$ denotes the integer part)

$$(\pi'_\theta(W(0, \beta))\,\psi)(\phi) = \psi(\phi + \beta \bmod 2\pi)\, e^{i\theta[(\phi+\beta)/2\pi]},$$

corresponding to $p_\theta = -i\partial/\partial\phi$ on quasi periodic functions satisfying the boundary condition $\psi(2\pi) = \psi(0)e^{i\theta}$. This representation corresponds to the self-adjoint extension which arises starting from $-i\partial/\partial\phi$ on C^∞ functions with support in $(0, 2\pi)$.

The representations of $\mathcal{A}$ can thus be seen to arise from those of $\mathcal{A}_{per}$ through the lifting of S^1 to its universal covering space $\mathbf{R}$, which involves the first homotopy group $\pi_1 = \mathbf{Z}$ of the circle. [48] A point of $x \in \mathbf{R}$ is in fact identified by a point of the circle ϕ and an integer n, called *winding number*, $x = \phi + 2\pi n$, $n = [x/2\pi]$.

One of the interesting features of this model is the interplay between topology and quantization, in connection with the occurrence of a nontrivial centre of the observable algebra. Such a feature is present in other more realistic models like electrons in a periodic potentials (the so-called Bloch electrons) [49] and, in our opinion, characterizes the vacuum structure

---

[48]See e.g. I.M. Singer and J.A. Thorpe, *Lecture Notes on Elementary Topology and Geometry*, Scott, Foresman and Company 1967.

[49]See J. Löffelholz, G. Morchio and F. Strocchi, Lett. Math. Phys. **35**, 251 (1995).

of Quantum Chromodynamics, where the centre is represented by the so-called large gauge transformations. [50]

3. **The functional integral.** We shall now discuss the role of the topology of the manifold on the corresponding stochastic process. For this purpose, as in Sect. 6.3 we consider the kernel of $\exp - H\tau$. Now, in each irreducible representation π_θ, the dynamics of a free particle on a circle is described by the (self-adjoint) Hamiltonian $H_\theta = p_\theta^2/(2mR^2)$. The spectrum of H_θ is discrete with eigenfunctions

$$\psi_n(\phi) = (2\pi)^{-1/2} e^{in\phi} e^{i\tilde{\theta}\phi}, \quad \tilde{\theta} \equiv \theta/(2\pi),$$

corresponding to the eigenvalues

$$E_n^\theta = E(n + \tilde{\theta})^2, \quad E \equiv (2mR^2)^{-1}, \quad n = 0, \pm 1, \ldots.$$

Now, we can easily compute the kernel of $\exp(-H_\theta\tau)$ by the spectral representation theorem in terms of (the kernel of) the spectral projections $P_n(\phi, \phi')$

$$K_\theta(\phi, \phi'; \tau, 0) = \sum_n P_n(\phi, \phi') e^{-E_n^\theta(\phi')\tau} = \sum_n \psi_n(\phi')^* \psi_n(\phi) e^{-E_n^\theta(\phi')\tau}$$

$$= (2\pi)^{-1} e^{i\tilde{\theta}\Delta\phi - E\tau\tilde{\theta}^2} \sum_n e^{-E\tau n^2 + n(i\Delta\phi - 2E\tilde{\theta}\tau)},$$

where $\Delta\phi \equiv \phi - \phi'$.

By using the Poisson summation formula [51]

$$\sum_{m=-\infty}^{m=\infty} e^{-Am^2 + 2\pi imz} = (\pi/A)^{1/2} \sum_{n=-\infty}^{n=\infty} e^{-(2\pi(z+n))^2/4A}, \tag{6.8.3}$$

---

[50] For the role of topology in QCD see R. Jackiw in S.B. Treiman, R. Jackiw, B. Zumino and E. Witten, *Current Algebra and Anomalies*, World Scientific 1985, esp. pp. 253-267; S. Coleman, *Aspects of Symmetry*, Cambridge Univ. Press 1985. The model mimics Quantum Chromodynamics without fermions and in this analogy the automorphisms of eq. (6.8.1) correspond to the "chiral" transformations induced by the topological charge. For the relation with gauge theories and the role of the center of the observables see J. Löffelholz, G. Morchio and F. Strocchi, Ann. Phys. **250**, 367 (1996).

[51] See e.g. P. Henrici, *Applied and Computational Complex Analysis*, Vol. 2, Wiley 1977, Sect. 10.6 IV. Given a function $f \in L^1(\mathbf{R})$ one can construct its periodization (of period 1) $f_{per}(x) = \sum_m f(x+m)$ (the series being convergent in the norm of $L^1([0, 1)))$, whose Fourier coefficients are

$$c_n = \int_0^1 dx\, f_{per}(x) e^{-2\pi inx} = \int_{-\infty}^\infty dx\, f(x) e^{-2\pi inx} = \tilde{f}(2\pi n).$$

Thus, one gets the *Poisson summation formula*

$$\sum_m f(x + m) = f_{per}(x) = \sum_n \int dy f(y)\, e^{-2\pi in(y-x)}.$$

By evaluating it at the point $x = 0$, for the function $f(y) = e^{-Ay^2 + 2\pi iyz}$, and by performing a Gaussian integration, one gets eq. (6.8.3).

the expression for K_θ can be cast in a more convenient form

$$K_\theta(\phi, \phi'; \tau, 0) = (4\pi E\tau)^{-1/2} \sum_n e^{-in\theta} e^{-(\phi - \phi' + 2\pi n)^2/(4E\tau)} \qquad (6.8.4)$$

$$\equiv \sum_n e^{-in\theta} K_n(\phi, \phi'; \tau, 0).$$

The kernel $K_{\theta=0}$ is positive and satisfies the semigroup property, so that by the same argument of Appendix J, there is a functional measure $d\mu_{\phi,\phi',\tau}(\phi(\cdot))$ on the path space $(S^1)^T, T = [0, \infty]$ for trajectories starting from ϕ' at time $t = 0$ and ending at ϕ at time $t = \tau$, such that

$$K_{\theta=0}(\phi, \phi'; \tau, 0) = \int d\mu_{\phi,\phi',\tau}(\phi(\cdot)). \qquad (6.8.5)$$

The so defined measure is closely related to the Wiener measure. In fact, the kernel K_n, defined in eq. (6.8.3), has the same form of the heat kernel K (the kernel $K_{\theta=0}$ is the 2π periodization of K) and can be given the meaning of probability that a trajectory starting from ϕ' reach the point ϕ at time τ after n crossings of the origin (counted algebraically, e.g. positive/negative if clockwise/anticlockwise). In this way the winding number enters in characterizing the trajectories and becomes a stochastic variable. Thus, we can write

$$\int d\mu_{\phi,\phi',\tau}(\phi(\cdot)) = \sum_n \int dW_{x=\phi+2\pi n, x'=\phi'}(x(\cdot)), \qquad (6.8.6)$$

where dW is the Wiener measure on trajectories $x(\tau) \in \mathbf{R}^T = (S^1 \times \mathbf{Z})^T$ starting from ϕ' and ending at $\phi + 2\pi n$. The above functional integral representation extends to kernels of products of $e^{-H\tau_j}$ and multiplication operators $A(\phi_j)$, as in the case of a quantum particle on a line (Sect. 6.3, eq. (6.3.1)).

4. The topological term. The case of $\theta \neq 0$ is slightly more intriguing since the kernel K_θ is not positive and therefore it cannot define a positive functional measure, nor have an interpretation in terms of a probability measure. To simplify the discussion, we consider the unitarily equivalent kernel

$$L_\theta(\phi, \phi'; \tau, 0) \equiv e^{-i\tilde{\theta}\phi} K_\theta(\phi, \phi'; \tau, 0) e^{i\tilde{\theta}\phi'}. \qquad (6.8.7)$$

Now, by repeating the same steps of the derivation of the Feynman-Kac formula in Sect. 6.3 and taking into account the periodicity of L_θ in the variables ϕ, ϕ', one gets the following representation for it

$$\lim_{k \to \infty} (N_k)^k \sum_n \int_{-\infty}^{\infty} dx_1 ... dx_{k-1} e^{-S_k(\phi+2\pi n, x_{k-1}, ..., x_1, \phi') - i\tilde{\theta}\Delta_n},$$

where $N_k \equiv (kmR^2/2\pi\tau)^{1/2}$, S_k is the k-th order finite difference euclidean action as in Sect. 6.3, with $x_0 = \phi'$, $x = \phi + 2\pi n$, and $\Delta_n \equiv \phi + 2\pi n - \phi'$. The sum

$$S_k^\theta \equiv S_k + i\tilde{\theta}\Delta_n$$

can be interpreted as the k-th order polygonal approximation of the following classical euclidean action

$$S^\theta = (m/2) \int_0^\tau \dot{x}^2(s)\, ds + i\tilde{\theta} \int_0^\tau \dot{x}(s)\, ds, \qquad (6.8.8)$$

which differs from the standard free action by the so-called *topological term*, proportional to θ.

A few remarks may be worthwhile. First, the non-trivial topology of the circle has led to a modification of the heuristic Feynman prescription, since the classical Lagrangean of a classical particle on a circle does not contain any θ parameter. This parameter appears as a hidden parameter, which shows up at the quantum level and it is related to the existence of inequivalent representations of the observable algebra.

The role of the topological term may look puzzling , since a total derivative in the classical Lagrangean has no effect on the equations of motion, and in fact it does not affect the euclidean equations. However, the construction of the momentum from the abelian euclidean algebra generated by $\phi(\tau)$ involves the Hamiltonian and the topological term changes the definition of the momentum conjugated to ϕ.

In terms of correlation functions, the role of the topological term is that of changing the boundary conditions for the wave functions at times $-T, T$, (eq. (6.5.7), i.e. $\Psi_0(x) \to e^{i\tilde{\theta}x}\Psi_0(x)$). In the thermodynamical limit $T \to \infty$, one goes from the correlation functions of the ground state of $H_{\theta=0}$ to those of the ground state of H_θ. [52] This should not be too surprising, since the topology is related to global properties of the manifold (which are not seen locally) and global properties of the dynamics are encoded in the ground state.

Finally, the topological term is complex and this explains why the corresponding functional measure is complex. This means that Nelson positivity fails, but one can show that Osterwalder-Schrader positivity is satisfied; this feature is shared by gauge quantum field theory models (see the references of the previous footnote).

5. The charged field algebra. Charged sectors. To appreciate the role of the non-trivial topology of the manifold (and also the relation between this model and gauge quantum field theory), it is convenient to extend the algebra $\mathcal{A}$ by embedding it into the standard Weyl algebra $\mathcal{A}_W$,

---

[52] J. Löffelholz, G. Morchio and F. Strocchi, Lett. Math. Phys. **35**, 251 (1995); Ann. Phys. **250**, 367 (1996).

(generated by the $W(\alpha, \beta)$, $\alpha, \beta \in \mathbf{R}$, eq. (3.2.1)), hereafter called *field al-gebra*. This corresponds to introducing the "decompactified" variable q, such that $\phi = q \mod 2\pi$, equivalently, to embedding the circle in the line. Such an extension of $\mathcal{A}$ naturally leads to the concept of a *gauge* group G of automorphisms, γ_m, $m \in \mathbf{Z}$ of the field algebra, with the property that they leave $\mathcal{A}$ pointwise invariant. They have the following form [53]

$$\gamma_m(W(\alpha, \beta)) = e^{i2\pi m\alpha} W(\alpha, \beta), \quad m \in \mathbf{Z}, \qquad (6.8.9)$$

and can be obtained by the adjoint action of the elements $W(0, 2\pi m)$ of the centre of $\mathcal{A}$.

Here, the gauge group G has the meaning of the group of translations of the "decompactified" variable q by $2\pi m$, i.e. $q \to q + 2\pi m$, and correspond to rotations on the circle of angle $2\pi m$. The gauge transformations are labeled by the winding number m related to the first homotopy group of the circle. In the analogy with gauge quantum field theory, they correspond to the so-called *large gauge transformations* of QCD. One of the interesting feature of the model is to display the relation between the origin of the gauge group and the non-trivial topology of the manifold.

The elements of the field algebra which are not in the observable al-gebra are called *charged fields*, since they are the analog of the charged fields in gauge quantum field theory and $\mathcal{A}_W$ plays the role of the field algebra extension of the observable algebra . The adjoint action of $W(\tilde{\theta}, 0)$ on $W(n, \beta)$ gives the automorphisms ρ^θ, defined by eq. (6.8.1), (therefore called *charged automorphisms*):

$$W(\tilde{\theta}, 0)\, W(n, \beta) W(-\tilde{\theta}, 0) = \rho^\theta(W(n, \beta)).$$

Actually, the extension from $\mathcal{A}$ to $\mathcal{A}_W$ can be seen as the minimal exten-sion which includes operators which intertwine between inequivalent rep-resentations of the observable algebra (called *charged sectors* in analogy with quantum field theory) and unitarily implement the automorphisms of eq. (6.8.1).

The representation $\pi_{\theta=0}$ can be obtained as the GNS representation defined by the ground state $\Omega_{\theta=0}$ of $\mathcal{H}_{\theta=0}$

$$\Omega_{\theta=0}(W(n, \beta)) = \delta_{n,0} = (\Psi_0, W(n, \beta)\,\Psi_0),$$

$\Psi_0(\phi) = (2\pi)^{-1/2} \in L^2([0, 2\pi), d\phi)$. The state

$$\Omega_\theta(W(n, \beta)) \equiv \Omega_{\theta=0}(\rho^\theta(W(n, \beta))) = e^{i\tilde{\theta}\beta}\, \delta_{n,0}$$

[53] For the general mathematical structure underlying this construction see F. Acerbi, G. Morchio and F. Strocchi, Lett. Math. Phys. **27**, 1 (1993); Jour. Math. Phys., **34**, 899 (1993).

defines the representation π_θ, through the GNS construction. The corresponding cyclic vector in $\mathcal{H}_\theta$ is $\Psi_\theta = (2\pi)^{-1/2}\, e^{i\tilde\theta\phi}$, as can be seen by using eq. (6.8.1).

It is worthwhile to remark that, even if both $\mathcal{H}_{\theta=0}$ and $\mathcal{H}_\theta$ are isomorphic to $L^2([0,2\pi), d\phi)$, there is no vector Φ_θ in $\mathcal{H}_{\theta=0}$ such that

$$(\Phi_\theta, \pi_\theta(W(n,\beta))\,\Phi_\theta) = e^{i\tilde\theta\beta}\,\delta_{n,0},$$

i.e. the two representations are indeed inequivalent. [54] Thus, the automorphisms ρ^θ which have the meaning of algebraic symmetries cannot be implemented by unitary operators in any π_θ, i.e., as one says, they are *spontaneously broken symmetries*; they are instead unitarily implemented, in each (irreducible) representation π of the field algebra, by $\pi(W(\tilde\theta,0))$. This technical advantage is at the basis of the standard strategy in the quantization of gauge quantum field theories in terms of a field algebra which includes charged fields.

6. Non regular representation of the Weyl algebra. As we have seen in Sect. 3, the irreducible regular representations of the field algebra $\mathcal{A}_W$ are given, up to unitary equivalence, by the standard Schroedinger representation π_S in $\mathcal{H} = L^2(\mathbf{R}, dx)$. This is a reducible representation of the observable algebra $\mathcal{A}$, and in fact $\pi_S(V(2\pi))$ commutes with all the observables and (therefore) defines a superselection rule. [55] Thus, not any (bounded) self-adjoint operator in $\mathcal{H}$ describes an observable and one of the Dirac-Von Neumann axioms for quantum mechanics fails (clearly the algebraic approach discussed in Sects. 2,3 does not have such a problem and actually provides the right way to clarify it). [56] The Hilbert space $\mathcal{H} = L^2(\mathbf{R}, dx)$ decomposes as a direct integral

$$\mathcal{H} = \int_0^{2\pi} d\theta\, \mathcal{H}_\theta,$$

over the spectrum of $\pi_S(V(2\pi))$ and one recovers in this way the irreducible representations $\mathcal{H}_\theta$ of $\mathcal{A}$; in particular the ground states $\Psi_\theta \in \mathcal{H}_\theta$ correspond to improper eigenvectors of $\pi_S(V(2\pi))$, since the spectrum of $\pi_S(V(2\pi))$ is continuous.

The requirement that the (irreducible) representation π of the field algebra decomposes as a direct sum of the irreducible representations of the observable algebra, i.e. that $W(0,\beta)$ is regularly represented and that the

---

[54] This can also be seen by noting that $\Omega_\theta(W(0,2\pi)) = e^{i\theta}$, whereas for *any* state $\Psi \in \mathcal{H}_{\theta=0}$, $(\Psi,\, \pi_0(e^{i2\pi p})\,\Psi) = 1$.

[55] This implies that there is no observable with non vanishing matrix elements between vectors with different spectral support relative to $V(2\pi)$ and therefore one cannot observe or prepare a coherent superposition of such vectors.

[56] For a beautiful review on superselection rules in quantum mechanics see A.S. Wightman, Nuovo Cim. **110 B**, 751 (1995).

spectrum of $\pi(W(0, 2\pi))$ is a pure point spectrum, so that the ground states of the Hamiltonians H_θ exist as proper vectors of $\mathcal{H}_\pi$, selects a unique irreducible representation π of the field algebra in which $W(\alpha, 0)$ is non-regularly represented. [57] In fact, if $e^{i\theta}$ is a point of the spectrum of $\pi(V(2\pi))$ and ω_θ is the corresponding eigenstate

$$e^{i\theta}\,\omega_\theta(U(\alpha)) = \omega_\theta(U(\alpha)V(2\pi))$$

$$= \omega_\theta(V(2\pi)\,U(\alpha)\,V(-2\pi)\,V(2\pi)) = e^{i(2\pi\alpha+\theta)}\,\omega_\theta(U(\alpha)),$$

so that $\omega_\theta(U(\alpha)) = 0$ if $\alpha \notin \mathbf{Z}$ and one cannot have strong continuity in α. Moreover, if P_0 denotes the projection on the eigenspace of $\pi(V(2\pi))$ with eigenvalue one, then $P_0\,\pi(V(\beta))\,P_0$ is a periodic function of period 2π and the strong continuity of $\pi(V(\beta))$ allows an easy extension of the argument for the representations of $\mathcal{A}_{per}$ to prove that $\mathcal{H}_\pi$ contains a vector Ψ_0 such that

$$(\Psi_0, \pi(W(\alpha, \beta))\,\Psi_0) = \delta_{\alpha,0}.$$

The GNS representation defined by such a state [58] contains the dense subspace D_0 of vectors of the form $\Psi = \sum_n a_n\pi(U(\alpha_n))\,\Psi_0$, where the sum runs over a finite set, with scalar product

$$(\Psi,\,\Phi) = \sum_{n,m}\bar{a}_n\,b_m\omega_0(U(-\alpha_n)\,U(\beta_m)) = \sum_n \bar{a}_n b_n.$$

D_0 may be represented by almost periodic functions $\psi(x) = \sum_n a_n e^{i\alpha_n x}$, $x \in \mathbf{R}$ with scalar product given by the ergodic mean

$$(\psi,\,\phi) = \lim_{L\to\infty}(2L)^{-1}\int_{-L}^{L} dx\,\overline{\psi}(x)\,\phi(x).$$

Finally the Hilbert space $\mathcal{H}_\pi$ is obtained by closing D_0 and can be identified with the space of formal sums

$$\sum_n a_n e^{i\alpha_n x},\ \{a_n\} \in l^2.$$

It is not difficult to recognize that $\mathcal{H}_\pi$ does indeed decompose as the direct sum over $\theta \in [0, 2\pi)$ of the spaces $\mathcal{H}_\theta$ discussed before. One can show that the Hamiltonian $H = p^2/2m$ can be obtained as a strong limit (on a dense domain) of elements of the observable algebra, that it has a unique ground state given by Ψ_0 and that it leaves each $\mathcal{H}_\theta$ invariant. One can also develop a functional integral approach to the ground state correlation functions of the fields in terms of a Wiener like measure and an

[57]F. Acerbi, G. Morchio and F. Strocchi, Jour. Math. Phys. **34**, 899 (1993); J. Löffelholz, G. Morchio and F. Strocchi, Lett. Math. Phys. **35**, 251 (1995).
[58]See J. Löffelholz, G. Morchio and F. Strocchi, Lett. Math. Phys. **35**, 251 (1995).

ergodic mean. The different representations of the observable algebra are defined by the states $\Psi_\theta = (2\pi)^{-1/2}\, e^{i\hat\theta x}$, which are the ground states of the restriction of the Hamiltonian to the subspaces $\mathcal{H}_\theta$. Such representations can be obtained as the thermodynamical limit of correlation functions as discussed in Sect. 6.5, eq. (6.5.7), by suitably choosing the boundary condition wave functions at times $-T$, T. The topological term has the effect of changing such boundary conditions. For a more expanded discussion see the reference of the last footnote.

6.9   Appendix H: The central limit theorem

Theorem 6.9.1 *(Central limit) Let $\{Y_j\}, j = 1, ..., n$, be a sequence of independent identically distributed random variables with zero mean and with finite variance σ. Then, the probability distribution of the normalized sum*

$$S_n(t) = \frac{1}{\sqrt{n}} \left(\sum_{j=1}^{n} Y_j \right),$$

converges weakly (i.e. as probability measure), as $n \to \infty$, to a Gaussian distribution with zero mean and variance σ.

Proof. For the proof one exploits the fact that the probability distribution $d\mu_f(x)$ of a random variable f is fully characterized by its Fourier transform, called its *characteristic function*,

$$c_f(t) = \int e^{itx} \, d\mu_f(x) = < e^{itf} >, \quad t \in \mathbf{R}. \tag{6.9.1}$$

Thus, by the independence and equal distribution of the Y_j's

$$< e^{itS_n} > = \prod_{j=1}^{n} < \exp i(t/\sqrt{n})Y_j > = < \exp i(t/\sqrt{n})Y_1 >^n = c(t/\sqrt{n})^n,$$

where $c(t) \equiv c_{Y_1}(t)$. Furthermore, since $< Y_1 >= 0$ and by the dominated convergence theorem one may interchange the limit with the integral, one has

$$\lim_{t \to 0} \int (e^{itx} - 1)t^{-2} \, d\mu_{Y_1}(x) = -(1/2) \int x^2 \, d\mu_{Y_1}(x) = -\sigma/2,$$

i.e.

$$(c(t) - 1)/t^2 \approx_{t \to 0} -\sigma/2.$$

Hence, $c(t) = 1 - \sigma t^2/2 + o(t^2)$ and

$$c(t/\sqrt{n})^n = (1 - t^2\sigma/2n + o(1/n))^n \to e^{-t^2\sigma/2}.$$

This proves the weak convergence stated in the Theorem.

6.10   Appendix I: Gaussian variables. Wick's theorem

A *Gaussian random variable* $f : \Omega \rightarrow \mathbf{R}$ with mean μ and variance σ^2 is characterized by the following distribution function

$$d\mu_f(x) = (2\pi\sigma^2)^{-1/2} \, \exp\left[-(x-\mu)^2/2\sigma^2\right] dx. \qquad (6.10.1)$$

For $\sigma^2 \rightarrow 0$ it reduces to $\delta(x-\mu)$. The Gaussian distribution is also called normal and denoted by $N(\mu, \sigma^2)$. Henceforth, for simplicity, unless specified otherwise, we shall always consider Gaussian variables of zero mean, $\mu = 0$.

The Fourier transform of $d\mu_f$ or characteristic function is given by $c_f(t) = \exp\left(-t^2\sigma^2/2\right)$ and in terms of it one can easily compute the n-th moments of f

$$< f^n > = \int x^n \, d\mu_f(x) \equiv < x^n >, \quad < x^{2n+1} > = 0,$$

$$< x^{2n} > = (-i)^{2n} (d/dt)^{2n} c_f(t)|_{t=0} = ((2n)!/n!)\sigma^{2n}/2^n.$$

A set $f_1, ..., f_n$ of random variables is called *jointly Gaussian* if their joint characteristic function is

$$c_{f_1,...,f_n}(t_1, ..., t_n) \equiv < \exp\left(i \sum t_i f_i\right) > = \exp\left[-(1/2) \sum_{ij} A_{ij} \, t_i t_j\right].$$
$$(6.10.2)$$

with A an $n \times n$ symmetric real positive definite matrix, $< A_{ij} > = < f_i f_j >$. If $M \equiv A^{-1}$ one has [59]

$$d\mu_{f_1,...,f_n}(x_1, ..., x_n) = (2\pi)^{-n/2} (\det A)^{-1/2} \exp\left[-(1/2) \sum_{ij} M_{ij} \, x_i x_j\right],$$
$$(6.10.3)$$

The computations with Gaussian variables are made easier by the use of Wick's theorem [60]

---

[59] We recall that for any $n \times n$ complex symmetric matrix A, with $[A, A^*] = 0$, $A + A^* \geq 0$ and with n non-zero eigenvalues λ_i, and any (complex) vector a_i, $(a, x) \equiv \sum a_i x_i$

$$\int dx_1...dx_n \exp\left[-(1/2)(x, Ax) + (a, x)\right] = (2\pi)^{n/2}(\det A)^{-1/2} e^{(a, A^{-1}a)/2},$$

since a change of variables $x = A^{-1}a + y$ and the diagonalization of A by an orthogonal matrix lead to a product of elementary Gaussian integrals.

[60] G.C. Wick, Phys. Rev. **80**, 268 (1950). For a discussion in the context of Gaussian processes see B. Simon, *The $P(\phi)_2$ Euclidean (Quantum) Field Theory*, Princeton University Press 1971, Chap. I.

Theorem 6.10.1 *(Wick) If $f_1, ..., f_n$ are jointly (not necessarily distinct) Gaussian random variables, then*

$$< f_1...f_{2k} >= \sum_{pairs} < f_{i_1} f_{j_1} > ... < f_{i_k} f_{j_k} >, \quad < f_1...f_{2k+1} >= 0,$$

(6.10.4)

where the sum is over all the $(2k)!/(2^k k!)$ ways of writing the product $f_1...f_{2k}$ as a product of k unordered (distinct) pairs.

Proof. First we note that the random variables $f_1..f_j$ are jointly Gaussian iff $\forall a = (a_1, ...a_n) \in \mathbf{R}^n$, $f = \sum_{i=1}^{j} a_i f_i$ is Gaussian. This follows from the fact that if $f_1..f_j$ are jointly Gaussian the characteristic function of f is given by

$$c_f(t) =< e^{it \sum a_j f_j} >= c_{f_1...f_n}(ta_1, ..., ta_n) = e^{-(1/2) \sum A_{ij} a_i a_j};$$

conversely, if f is Gaussian

$$c_{f_1...f_n}(t_1, ..., t_n) \equiv e^{i \sum t_i f_i} = e^{-<f^2>/2} = e^{-\sum t_i t_j A_{ij}/2},$$

with $A_{ij} \equiv < f_i f_j >$.

The equations of the Theorem can then be obtained by taking suitable derivatives of the joint characteristic function at $t = 0$.

Another useful tool for computations is the notion of *Wick n-th power* $: f^n :$ of a Gaussian random variable. It is recursively defined by

$$: f^0 := 1, \quad : f^n := f : f^{n-1} : -(n-1) < f^2 > : f^{n-2} : . \quad (6.10.5)$$

Similarly, one defines the Wick exponential as an L^1 convergent series of Wick powers

$$: \exp \lambda f := \sum_{n=0}^{\infty} (\lambda^n/n!) : f^n :, \quad \lambda \in \mathbf{R}.$$

By differentiating order by order one can easily prove that

$$: \exp \lambda f := \exp \lambda f / < \exp \lambda f >= \exp(-\lambda^2 < f^2 > /2) \exp \lambda f.$$

An easy consequence of this equation is that for any two random variables f_1, f_2,

$$<: \exp \lambda_1 f_1 :: \exp \lambda_2 f_2 :>= \exp(\lambda_1 \lambda_2 < f_1 f_2 >),$$

where we have used that $\lambda_1 f_1 + \lambda_2 f_2$ is a Gaussian variable. By differentiation one also gets

$$<: f_1^n :: f_2^m :>= \delta_{nm} n! < f_1 f_2 >^n .$$

6.11   Appendix J: Stochastic processes and functional integrals

A *stochastic process* is a family of random variables, $\{f_t, \ t \in T\}$, defined on the same probability space (Ω, Σ, μ), where T is some index set. This implies that for any finite set $t_1, \dots t_n \in T$, one has joint probability distributions

$$d\mu_{f_{t_1}, \dots, f_{t_n}}(x_1, \dots, x_n) \equiv d\mu_{t_1, \dots, t_n}(x_1, \dots, x_n) \equiv P_{t_1, \dots, t_n}(x_1, \dots, x_n)dx_1 \dots dx_n,$$

(in the last notation P may be a Schwartz distribution), which satisfy the compatibility conditions:

1) $d\mu_{t_1, \dots, t_n}(x_1, \dots, x_n)$ is a positive measure
2) $\int d\mu_{t_1, \dots, t_n} = 1$
3) $\int P_{t_1, \dots, t_n}(x_1, \dots, x_n)\, dx_1 = P_{t_2, \dots, t_n}(x_2, \dots, x_n).$

A process is called *Gaussian* if the finite dimensional measures are of the form of eq. (6.10.3).

Whereas it is relatively simple to construct stochastic processes when the index set is finite or even denumerable, structural problems arise when T is not denumerable (and f_t is at least a Schwartz distribution in t) [61]. The basic result in this direction is Kolmogorov theorem, by which the knowledge of the joint probability distributions $d\mu_{t_1, \dots, t_n}(x_1, \dots, x_n)$ for any finite set $t_1, \dots, t_n$, satisfying 1)-3), uniquely determine (up to isomorphisms) the stochastic process $\{f_t, t \in T\}$.

The theorem does not only solve an existence problem, but it also sheds light on the concept of stochastic process and it establishes a deep connection between stochastic processes and functional integrals, i.e. *measures on infinite-dimensional spaces* of paths or trajectories.

For this purpose, we note that given a family $\{f_t, t \in T\}$ of random variables, defined on the same probability space (Ω, Σ, μ), also called *basic space*, for each $\omega \in \Omega$, $f_t(\omega)$ defines a function of t, i.e. a *path* or *trajectory* associated with the point ω. As it is standard in the theory of random variables, one can view a random variable $f : \Omega \to \mathbf{R}^n$ as a variable $x = \{x_j, j = 1, \dots, n\} \in \mathbf{R}^n$, (coordinate space picture), with probability distribution $d\mu_f(x)$ induced by f and μ on $\mathbf{R}^n$, with the natural mapping of Ω into $\mathbf{R}^n$ and of Σ in the Borel σ-field of $\mathbf{R}^n$, (see Sect. 2.4). The case of a stochastic process $\{f_t, t \in T\}$ with T infinite dimensional and, for simplicity, $f_t \in \mathbf{R}$, can be regarded as the case in which the coordinate space $\mathbf{R}^n$ becomes infinite-dimensional with cardinality T, namely $\mathbf{R}^T$, and a point of such a space can be viewed as a path $\{x_t, t \in T\}$.

---

[61]See J.L. Doob, *Stochastic Processes*, Wiley 1953, Ch. II; for a lucid brief account see J.L. Doob, Bull. AMS **53**, 15 (1947).

Following Nelson, in order to treat the case in which the path may pass through infinity, it is convenient to consider the one-point compactification $\dot{\mathbf{R}} = \mathbf{R} \cup \{\infty\}$ of $\mathbf{R}$, and $\dot{\mathbf{R}}^T$ as "coordinate" space or path space. This is not a restriction if the finite dimensional probability distributions vanish when one of the variables goes to infinity (or more generally give the same weight to $\pm\infty$); on the other hand, this choice is technically very useful because, by Tychonoff theorem,[62] $\dot{\mathbf{R}}^T$ with the product topology is a compact Hausdorff space.

The standard version of Kolmogorov theorem constructs the measure on path space starting from the so-called *cylinder sets*,

$$C_{t_1,\ldots,t_n;A} = \{x \in \dot{\mathbf{R}}^T ; (x_{t_1}, \ldots, x_{t_n}) \in A\},$$

with A a Borel set of $\mathbf{R}^n$, where only a finite set of "coordinates" are constrained, all the others being let free. The finite dimensional probability distributions (satisfying the compatibility conditions 1)-3)) define the measure on the cylinder sets and by a general measure theoretical result due to Kolmogorov the so defined measure has a unique extension to the σ-algebra generated by the cylinder sets. In such a version, a crucial role is played by the cylinder sets: the σ-algebra generated by them coincides with the Baire σ-algebra (namely the minimal algebra needed for the measurability of the continuous functions) [63] which may be smaller than the σ-algebra of Borel sets (namely the algebra generated by all open sets) of $\dot{\mathbf{R}}^T$ and therefore be inadequate for the discussion of interesting functional integral problems.

A very elegant version of Kolmogorov theorem, which directly gets a regular Borel measure [64] on path space $X \equiv \dot{\mathbf{R}}^T$, is due to Nelson. [65] The idea is to resolve the ambiguity connected with the choice of the σ-algebra of sets by emphasizing the role of the continuous functions (in agreement with the logic discussed in Sects. 1.2, 1.3), in determining the probability measure on the basic space. The strategy is to translate the information carried by the finite dimensional probability distributions into a positive linear functional on the set $C_{fin}(X)$ of continuous functions of the trajectories, which depend only on the values taken at a finite number

---

[62] See e.g. the quoted book by Reed and Simon, Vol. 1, p. 100.

[63] For a compact Haussdorff space the Baire σ-algebra is that generated by the compact G_δ sets, namely by the (compact) countable intersections of open sets. For a simple discussion of the distinction between Baire and Borel sets see M. Reed and B. Simon, *Methods of Modern Mathematical Physics*, Vol. I, Academic Press 1973, Sect. IV.4. For the role of Baire and Borel σ-algebras in the context of stochastic processes see E. Nelson, Ann. Math. **69**, 630 (1959).

[64] In general, a Baire measure has many extensions to the Borel σ-algebra, but only one which is regular (see P.R. Halmos, *Measure Theory*, Springer 1974, Chap. X, Theor. H; D.L. Cohn, *Measure Theory*, Birkhäuser 1980, Chap. 7).

[65] E. Nelson, Ann. Math. **69**, 630 (1959); *Quantum Fluctuations*, Princeton University Press 1985, Sect. I.3.

of times, and then use the Riesz-Markov theorem to get a regular Borel measure on the path space.

More precisely, $F \in C_{fin}(X)$ if there exists a *finite* subset $T_0 = \{t_1, ..., t_n\}$ $\subseteq T$ such that $F(x(\cdot)) = F(x'(\cdot))$, whenever the two trajectories $x(t), x'(t)$ take the same values for all times $t \in T_0$, i.e. F is a function depending only on the values taken by the trajectories at the times $t_1, ..., t_n$,

$$F(x(\cdot)) = F(x(t_1), ..., x(t_n)).$$

Clearly, $C_{fin}(X) \ni \mathbf{1} \equiv F(x(\cdot)) = 1$. Then one may define a positive linear functional on $C_{fin}(X)$

$$L(F) \equiv \int_{\mathbf{R}^n} F(x_1, ..., x_n) \, d\mu_{t_1, ..., t_n}(x_1, ..., x_n), \quad L(\mathbf{1}) = 1,$$

which is well defined, thanks to the compatibility conditions 1-3.

Since $C_{fin}(X)$ contains the identity and separates the points of X, by the Stone-Weierstrass theorem it is dense in $C(X)$, in the sup norm. Then, L has a unique extension to a positive linear functional on $C(X)$ and by the Riesz-Markov theorem determines a unique regular probability measure ν on X, such that $\forall F \in C_{fin}(X)$

$$L(F) = \int_X F(x(t_1), ..., x(t_n)) \, d\nu.$$

In this way, one proves the existence of a regular functional measure on $\dot{\mathbf{R}}^T$ and its construction is unique starting from the finite dimensional probability distributions, thanks to the property of regularity.

6.12   Appendix K: Wiener process

For the convenience of the reader we briefly discuss the basic properties of the Brownian motion viewed as a Wiener stochastic process, i.e. in terms of a functional integral defined by the Wiener measure.

The position $x(t)$ of a Brownian particle, which at time $t = 0$ starts from $x_0 = 0$, is a stochastic variable with probability distribution given by the heat kernel $K(x, 0; t, 0)$ (see Sect. 6.1). For any n-ple $t_1, ... t_n$, by the semigroup property of K, the joint probability distributions given by *products* of K's

$$d\mu_{t_1,...,t_n}(x_1, ..., x_n) = K(x_n, x_{n-1}; t_n, t_{n-1})...K(x_2, x_1; t_2, t_1) K(x_1, 0; t_1, 0)$$

satisfy the compatibility conditions of Appendix J. Thus, they define a stochastic process $\{x(t), t \in [0, \infty)\}$, identified by a functional measure $dW_0(x(\cdot))$ on the space X of trajectories starting from $x(0) = 0$, with the following expectations (for simplicity we put $D = 1/2$)

$$< x(t_1) >= \int_X x(t_1) \, dW_0(x(\cdot)) = \int dx_1 \, x_1 \, K(x_1, 0; t_1, 0) = 0,$$

$$< x(t_1) \, x(t_2) >= \int dx_1 \, dx_2 \, x_1 \, x_2 \, K(x_2, x_1; t_2, t_1) K(x_1, 0; t_1, 0)$$

$$= \int dx_1 \, x_1^2 \, K(x_1, 0; t_1, 0) = t_1, \ 0 \le t_1 \le t_2,$$

and in general

$$< x(t_1) \, x(t_2) >= (-|t_1 - t_2| + |t_1| + |t_2|)/2 = \min(t_1, t_2). \qquad (6.12.1)$$

The process is Gaussian (see Sect. 6.1) and therefore all the higher moments are computed in terms of the first two (see Appendix I). (For this reason the Brownian motion is sometimes defined as a Gaussian process with $x_0 = 0$, zero mean and variance $\min(t_1, t_2)$.)

By the costruction discussed in Appendix J, one obtains in this way a functional measure dW_0, which is called the *Wiener measure* and the corresponding process is the *Wiener process* .

If one tries to use the above strategy to control the limit $n \to \infty$ in eq. (6.2.3) with $V = 0$, when the kernel is

$$G^\alpha(x', x; t', t) \equiv (2\pi(t' - t)\alpha)^{-1/2} \, e^{-(x'-x)^2/2\alpha \, (t'-t)},$$

with $\alpha \in \mathbf{C}$, $\operatorname{Re} \alpha \ge 0$, one finds that the polygonal approximation cannot define a σ-additive measure in the limit, unless $\operatorname{Im} \alpha = 0$. [66] The point is

---

[66] R.H. Cameron, Jour. Math and Phys. **39**, 126(1960); for the essence of the argument see M. Reed and B. Simon, Vol. II, Problem 64.

that

$$\lim_{n \to \infty} \int dx_1 ... dx_n \, G^\alpha(x_n, x_{n-1}; t/n) ... G^\alpha(x_2, x_1; t/n) = 1,$$

$$\int dx_1 ... dx_n \, |G^\alpha(x_n, x_{n-1}; t/n)| ... |G^\alpha(x_2, x_1; t/n)| \simeq (|\alpha|/\mathrm{Re}\,\alpha)^n \to \infty,$$

so that if a complex measure $d\mu$ existed when $\mathrm{Im}\,\alpha \neq 0$, it could not have finite variation, since $\int |d\mu| = \infty$ and it could not be σ-additive.[67]

From the variance one can derive the regularity properties of the trajectories on which the functional measure is concentrated.

First, the relevant trajectories can be chosen Hölder continuous of order α, for any $\alpha < 1/2$, and therefore continuous. [68] This follows from Kolmogorov regularity theorem according to which if $x_t, t \in \mathbf{R}$ is a stochastic process and

$$E(|x_t - x_s|^\beta) \equiv < |x_t - x_s|^\beta > \leq c \, |t - s|^{1+\gamma}, \ \gamma < \beta,$$

then the process can be realized in terms of trajectories which are α-Hölder continuous, for any $\alpha < \gamma/\beta$. [69] For the Brownian motion, since $|x_t - x_s|$ is a Gaussian variable, $< |x_t - x_s|^{2n} > = C_n < |x_t - x_s|^2 >^n$ and by eq.(6.12.1) and symmetry $< |x_t - x_s|^2 > = |t - s|$, so that one has Hölder continuity for any $\alpha < 1/2 - (2n)^{-1}, \forall n$.

Secondly, apart from a set of zero Wiener measure, the trajectories of the Brownian motion do not have continuous derivatives. A simple indication of this fact comes from

$$< |x_t - x_s|^2 / |t - s|^2 > = |t - s|^{-1}$$

so that the limit $t \to s$ is divergent. A more precise argument is obtained by using the following relations. A function $f : [a, b] \to \mathbf{R}$ is absolutely continuous if there exists a $g \in L^1(a, b)$ such that

$$f(x) = f(a) + \int_a^x g(y) \, dy.$$

This property is equivalent to the almost everywhere existence of the derivative $f' \in L^1$. Absolute continuity implies that f is of bounded variation, i.e.

$$\sup_\pi \sum_{i=1}^n |f(x_{i+1}) - f(x_i)| < \infty,$$

---

[67]See e.g. I.E. Segal and R.A. Kunze, *Integrals and Operators*, Springer 1978, Sect. 4.2.

[68]We recall that a function f is Hölder continuous of order α in a domain D if there is a constant c such that $|f(x) - f(y)| \leq c|x - y|^\alpha, \ \forall x, y \in D$.

[69]For the proof see e.g. B. Simon, *Functional Integration and Quantum Physics*, Academic Press 1979, Th. 5.1, p. 43.

where π is any partition $\{a = t_1 < ... < t_{n+1} = b\}$, since

$$\sum_i |f(x_{i+1}) - f(x_i)| \le \sum_i \int_{x_i}^{x_{i+1}} |g(y)| \, dy = \int_a^b |g(y)| \, dy.$$

Finally, if f is continuous and of bounded variation, then its quadratic variation in $[a, b]$

$$\Delta^{(2)} f \equiv \lim_{\sup_j |x_{j+1} - x_j| \to 0} \sum_i |f(x_{i+1}) - f(x_i)|^2$$

$$\le \sup_i |f(x_{i+1}) - f(x_i)| \sum_j |f(x_{j+1}) - f(x_j)|$$

vanishes, because f is uniformily continuous in $[a, b]$ and

$$\sum_i |f(x_{i+1}) - f(x_i)| < c$$

since f is of bounded variation.

Therefore, to prove that the relevant trajectories of the Brownian motion are not continously differentiable it suffices to prove that their quadratic variation in $[a, b]$ is almost surely non zero, i.e. that, for $t \in [a, b]$, $\Delta^{(2)} x = b - a$, in $L^2(X, dW)$. Indeed, $< \Delta^{(2)} x >= b - a$, and for any partition π, putting $|\pi| \equiv \sup_j |t_{j+1} - t_j|$,

$$< |\Delta_\pi^{(2)} f|^2 >=< |\sum_i |f(x_{i+1}) - f(x_i)|^2 \,|^2 >= (b-a)^2 + 2 \sum_i (t_{i+1} - t_i)^2$$

so that

$$< |\Delta_\pi^{(2)} f - (b-a)|^2 >\le 2|\pi|(b-a) \to 0$$

when $|\pi| \to 0$.

This shows that the Wiener measure is supported by trajectories which have a rather irregular behaviour and therefore, in dealing with the corresponding functional integral, the validity of approximations based on regular paths is problematic.

Index

Printed in the United States
By Bookmasters